Introduction to DIGITAL MICROELECTRONIC CIRCUITS

Introduction to
DIGITAL MICROELECTRONIC CIRCUITS

Er. N.B.L. Mathur

ANMOL PUBLICATIONS PVT. LTD.
NEW DELHI-110 002 (INDIA)

ANNOL PUBLICATIONS PVT. LTD.
Regd. Office: 4360/4, Ansari Road, Daryaganj,
New Delhi-110002 (India)
Tel.: 23278000, 23261597, 23286875, 23255577
Fax: 91-11-23280289
Email: anmolpub@gmail.com
Visit us at: www.anmolpublications.com

Branch Office: No. 1015, Ist Main Road, BSK IIIrd Stage
IIIrd Phase, IIIrd Block, Bangalore-560 085 (India)
Tel.: 080-41723429 • Fax: 080-26723604
Email: anmolpublicationsbangalore@gmail.com

Introduction to Digital Microelectronic Circuits

First Edition, 2010

ISBN 978-81-261-4462-4

PRINTED IN INDIA

Printed at Mehra Offset Press, Delhi.

Contents

Preface

The book explains thoroughly the implementation of logic gates using different configurations of MOS devices. Extending the analysis of digital IC families it cover the more recent BiCMOS and GaAs technologies. The book gives a balanced treatment of regenerative logic circuits using bipolar and MOS discrete and integrated circuits.

The book include coverage of popular methods of analog-digital data conversions, introduces LSI and VLSI systems with memories and gate arrays. The vast amount of material arising from innovative circuit designs and newer device technologies requires that the circuit analysis aspects of digital electronics be covered in a first course, separate from device design and chip layout. Consequently, Introduction to Digital Microelectronic Circuits emphasizes the analysis and performance comparison of different gate-level logic circuits and presents design examples based on logic-level requirements. It provides an introduction to the analysis of digital electronic circuits using discrete and integrated circuits.

Author

Chapter 1

Introduction

DIGITAL CIRCUITS

The quantities that are to be measured, monitored, recorded, processed and controlled are analog and digital, depending on the type of system used. It is important when dealing with various quantities that we be able to represent their values efficiently and accurately. There are basically two ways of representing the numerical value of quantities, analog and digital.

ANALOG REPRESENTATION

Systems which are capable of processing a continuous range of values varying with respect to time are called analog systems. In analog representation a quantity is represented by a voltage, current, or meter movement that is proportional to the value of that quantity. Analog quantities such as those cited above have an important characteristic: They can vary over a continuous range of values.

Analog Voltage vs Time

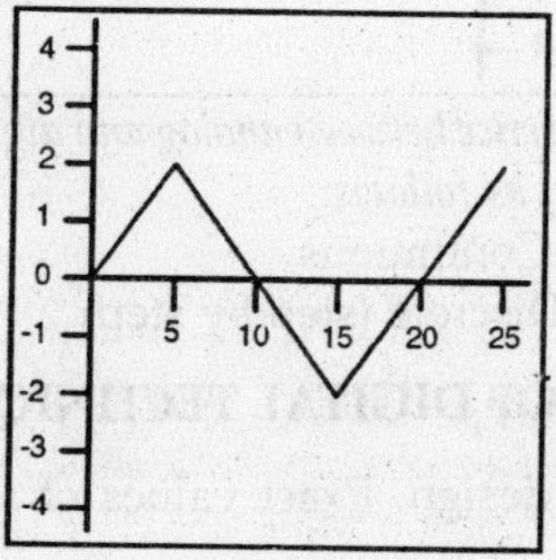

DIGITAL REPRESENTATION

Systems which process discrete values are called digital systems. In digital representation the quantities are represented not by proportional quantities but by symbols called digits.

As an example, consider the digital watch, which provides the time of the day in the form of decimal digits representing hours and minutes (and sometimes seconds). As we know, time of day changes continuously, but the digital watch reading does not change continuously; rather, it changes in steps of one per minute (or per second).

In other words, time of day digital representation changes in discrete steps, as compared to the representation of time provided by an analog watch, where the dial reading changes continuously. *Below is a diagram of digital voltage vs time*: Here input voltage changes from +4 Volts to -4 Volts; it can be converted to digital form by Analog to Digital converters (ADC). An ADC converts continuous signals into samples per second. Well, this is an entirely different theory.

Digital Voltage vs Time

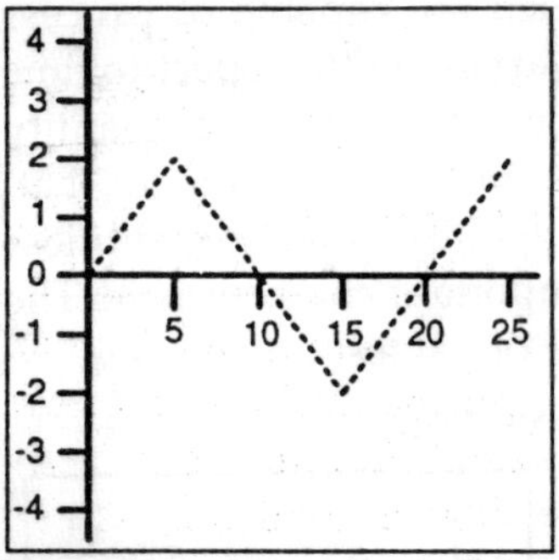

The major difference between analog and digital quantities, then, can be stated simply as follows:

- Analog = Continuous
- Digital = Discrete (step by step)

ADVANTAGES OF DIGITAL TECHNIQUES

- Easier to design. Exact values of voltage or current

are not important, only the range (HIGH or LOW) in which they fall.

- Information storage is easy.
- Accuracy and precision are greater.
- Operations can be programmed. Analog systems can also be programmed, but the available operations variety and complexity is severely limited.
- Digital circuits are less affected by noise, as long as the noise is not large enough to prevent us from distinguishing HIGH from LOW (we discuss this in detail in an advanced digital tutorial section).
- More digital circuitry can be fabricated on IC chips.

LIMITATIONS OF DIGITAL TECHNIQUES

- Most physical quantities in real world are analog in nature, and these quantities are often the inputs and outputs that are being monitored, operated on, and controlled by a system. Thus conversion to digital format and re-conversion to analog format is needed.

NUMBER SYSTEM

Many number systems are in use in digital technology. The most common are the decimal, binary, octal, and hexadecimal systems. The decimal system is clearly the most familiar to us because it is a tool that we use every day. Examining some of its characteristics will help us to better understand the other systems. In the next few pages we shall introduce fcur numerical representation systems that are used in the digital system. There are other systems, which we will look at briefly.

- Decimal
- Binary
- Octal
- Hexadecimal

DECIMAL SYSTEM

The decimal system is composed of 10 numerals or symbols. These 10 symbols are 0, 1, 2, 3, 4, 5, 6, 7, 8, 9. Using these symbols as digits of a number, we can express any quantity. The decimal system is also called the base-10 system because it has 10 digits.

10^3	10^2	10^1	10^0		10^{-1}	10^{-2}	10^{-3}
=1000	=100	=10	=1		=0.1	=0.01	=0.001
Most Significant Digit				Decimal point			Least Significant Digit

Even though the decimal system has only 10 symbols, any number of any magnitude can be expressed by using our system of positional weighting.

Decimal Examples

- 3.14_{10}
- 52_{10}
- 1024_{10}
- 64000_{10}

BINARY SYSTEM

In the binary system, there are only two symbols or possible digit values, 0 and 1. This base-2 system can be used to represent any quantity that can be represented in decimal or other base system.

2^3	2^2	2^1	2^0		2^{-1}	2^{-2}	2^{-3}
=8	=4	=2	=1		=0.5	=0.25	=0.125
Most Significant Digit				Binary point			Least Significant Digit

BINARY COUNTING

Table. Binary Counting Sequence

2^3	2^2	2^1	2^0	Decimal
0	0	0	0	0
0	0	0	1	1
0	0	1	0	2
0	0	1	1	3

0	1	0	0	4
0	1	0	1	5
0	1	1	0	6
0	1	1	1	7
1	0	0	0	8
1	0	0	1	9
1	0	1	0	10
1	0	1	1	11
1	1	0	0	12
1	1	0	1	13
1	1	1	0	14
1	1	1	1	15

Representing Binary Quantities

In digital systems the information that is being processed is usually presented in binary form. Binary quantities can be represented by any device that has only two operating states or possible conditions. E.g.. a switch is only open or closed. We arbitrarily (as we define them) let an open switch represent binary 0 and a closed switch represent binary 1. Thus we can represent any binary number by using series of switches.

Typical Voltage Assignment

Binary 1: Any voltage between 2V to 5V

Binary 0: Any voltage between 0V to 0.8V

Not used: Voltage between 0.8V to 2V in 5 Volt CMOS and TTL Logic, this may cause error in a digital circuit. Today's digital circuits works at 1.8 volts, so this statement may not hold true for all logic circuits.

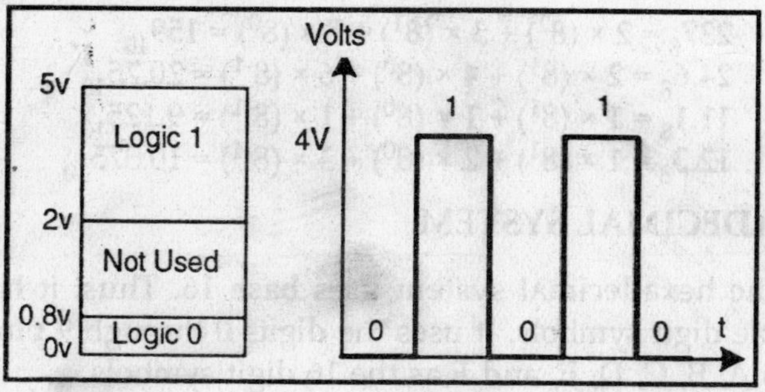

We can see another significant difference between digital and analog systems. In digital systems, the exact voltage value is not important; eg, a voltage of 3.6V means the same as a voltage of 4.3V. In analog systems, the exact voltage value is important.

The binary number system is the most important one in digital systems, but several others are also important. The decimal system is important because it is universally used to represent quantities outside a digital system. This means that there will be situations where decimal values have to be converted to binary values before they are entered into the digital system.

In additional to binary and decimal, two other number systems find wide-spread applications in digital systems. The octal (base-8) and hexadecimal (base-16) number systems are both used for the same purpose- to provide an efficient means for representing large binary system.

OCTAL SYSTEM

The octal number system has a base of eight, meaning that it has eight possible digits: 0,1,2,3,4,5,6,7.

8^3	8^2	8^1	8^0		8^{-1}	8^{-2}	8^{-3}
=512	=64	=8	=1	.	=1/8	=1/64	=1/512
Most Significant Digit				Octal point			Least Significant Digit

Octal to Decimal Conversion

- $237_8 = 2 \times (8^2) + 3 \times (8^1) + 7 \times (8^0) = 159_{10}$
- $24.6_8 = 2 \times (8^1) + 4 \times (8^0) + 6 \times (8^{-1}) = 20.75_{10}$
- $11.1_8 = 1 \times (8^1) + 1 \times (8^0) + 1 \times (8^{-1}) = 9.125_{10}$
- $12.3_8 = 1 \times (8^1) + 2 \times (8^0) + 3 \times (8^{-1}) = 10.375_{10}$

HEXADECIMAL SYSTEM

The hexadecimal system uses base 16. Thus, it has 16 possible digit symbols. It uses the digits 0 through 9 plus the letters A, B, C, D, E, and F as the 16 digit symbols.

16^3	16^2	16^1	16^0		16^{-1}	16^{-2}	16^{-3}
=4096	=256	=16	=1	.	=1/16	=1/256	=1/4096
Most Significant Digit				Hexa Decimal point			Least Significant Digit

Hexadecimal to Decimal Conversion

- $24.6_{16} = 2 \times (16^1) + 4 \times (16^0) + 6 \times (16^{-1}) = 36.375_{10}$
- $11.1_{16} = 1 \times (16^1) + 1 \times (16^0) + 1 \times (16^{-1}) = 17.0625_{10}$
- $12.3_{16} = 1 \times (16^1) + 2 \times (16^0) + 3 \times (16^{-1}) = 18.1875_{10}$

BINARY ARITHMETIC

NUMBERS VERSUS NUMERATION

It is imperative to understand that the type of numeration system used to represent numbers has no impact upon the outcome of any arithmetical function (addition, subtraction, multiplication, division, roots, powers, or logarithms). A number is a number is a number; one plus one will always equal two (so long as we're dealing with real numbers), no matter how you symbolize one, one, and two.

A prime number in decimal form is still prime if it's shown in binary form, or octal, or hexadecimal? is still the ratio between the circumference and diameter of a circle, no matter what symbol(s) you use to denote its value. The essential functions and interrelations of mathematics are unaffected by the particular system of symbols we might choose to represent quantities. This distinction between numbers and systems of numeration is critical to understand.

The essential distinction between the two is much like that between an object and the spoken word(s) we associate with it. A house is still a house regardless of whether we call it by its English name house or its Spanish name casa. The first is the actual thing, while the second is merely the symbol for the thing.

That being said, performing a simple arithmetic operation such as addition (longhand) in binary form can be confusing to a person accustomed to working with decimal numeration only. In this lesson, we'll explore the techniques used to perform simple arithmetic functions on binary numbers, since these techniques will be employed in the design of electronic circuits to do the same.

You might take longhand addition and subtraction for granted, having used a calculator for so long, but deep inside that calculator's circuitry all those operations are performed"longhand," using binary numeration. To understand how that's accomplished, we need to review to the basics of arithmetic.

BINARY ADDITION

Adding binary numbers is a very simple task, and very similar to the longhand addition of decimal numbers. As with decimal numbers, you start by adding the bits (digits) one column, or place weight, at a time, from right to left. Unlike decimal addition, there is little to memorize in the way of rules for the addition of binary bits:

0 + 0 = 0

1 + 0 = 1

0 + 1 = 1

1 + 1 = 10

1 + 1 + 1 = 11

Just as with decimal addition, when the sum in one column is a two-bit (two-digit) number, the least significant figure is written as part of the total sum and the most significant figure is "carried" to the next left column. Consider the following examples:

- 11 1← Carry bits → 11
- 1001101 1001001 1000111
- + 0010010 + 0011001 + 0010110
- ——— ———— ————
- 1011111 1100010 1011101

The addition problem on the left did not require any bits to be carried, since the sum of bits in each column was either 1 or 0, not 10 or 11. In the other two problems, there definitely were bits to be carried, but the process of addition is still quite simple. As we'll see, there are ways that electronic circuits can be built to perform this very task of addition, by representing each bit of each binary number as a voltage signal (either "high," for a 1; or "low" for a 0). This is the very foundation of all the arithmetic which modern digital computers perform.

NEGATIVE BINARY NUMBERS

With addition being easily accomplished, we can perform the operation of subtraction with the same technique simply by making one of the numbers negative.

For example, the subtraction problem of 7 – 5 is essentially the same as the addition problem 7 + (–5). Since we already know how to represent positive numbers in binary, all we need to know now is how to represent their negative counterparts and we'll be able to subtract.

Usually we represent a negative decimal number by placing a minus sign directly to the left of the most significant digit, just as in the example above, with -5.

However, the whole purpose of using binary notation is for constructing on/off circuits that can represent bit values in terms of voltage (2 alternative values: either "high" or "low").

In this context, we don't have the luxury of a third symbol such as a "minus" sign, since these circuits can only be on or off (two possible states). One solution is to reserve a bit (circuit) that does nothing but represent the mathematical sign:

- $101_2 = 5_{10}$ (positive)
- Extra bit, representing sign (0=positive, 1=negative)
- |
- $0101_2 = 5_{10}$ (positive)
- Extra bit, representing sign (0=positive, 1=negative)
- |
- $1101_2 = -5_{10}$ (negative)

As you can see, we have to be careful when we start using bits for any purpose other than standard place-weighted values. Otherwise, 1101_2 could be misinterpreted as the number thirteen when in fact we mean to represent negative five. To keep things straight here, we must first decide how many bits are going to be needed to represent the largest numbers we'll be dealing with, and then be sure not to exceed that bit field length in our arithmetic operations.

For the above example, I've limited myself to the representation of numbers from negative seven (1111_2) to positive seven (0111_2), and no more, by making the fourth bit the "sign" bit. Only by first establishing these limits can I avoid confusion of a negative number with a larger, positive number.

Representing negative five as 1101_2 is an example of the *sign-magnitude* system of negative binary numeration. By using the leftmost bit as a sign indicator and not a place-weighted value, I am sacrificing the "pure" form of binary notation for something that gives me a practical advantage: the representation of negative numbers.

The leftmost bit is read as the sign, either positive or negative, and the remaining bits are interpreted according to the standard binary notation: left to right, place weights in multiples of two.

As simple as the sign-magnitude approach is, it is not very practical for arithmetic purposes. For instance, how do I add a negative five (1101_2) to any other number, using the standard technique for binary addition? I'd have to invent a new way of doing addition in order for it to work, and if I do that, I might as well just do the job with longhand subtraction; there's no arithmetical advantage to using negative numbers to perform subtraction through addition if we have to do it with sign-magnitude numeration, and that was our goal!

There's another method for representing negative numbers which works with our familiar technique of longhand addition, and also happens to make more sense from a place-weighted numeration point of view, called *complementation*. With this strategy, we assign the leftmost bit to serve a special purpose, just as we did with the sign-magnitude approach, defining our number limits just as before. However, this time, the leftmost bit is more than just a sign bit; rather, it possesses a negative place-weight value.

For example, a value of negative five would be represented as such:

Extra bit, place weight = negative eight:

- |

- $1011_2 = 5_{10}$ (negative)
- $(1 \times -8_{10}) + (0 \times 4_{10}) + (1 \times 2_{10}) + (1 \times 1_{10}) = -5_{10}$

With the right three bits being able to represent a magnitude from zero through seven, and the leftmost bit representing either zero or negative eight, we can successfully represent any integer number from negative seven,

$(1001_2 = -8_{10} + 7_{10} = -1_{10})$

to positive seven

$(0111_2 = 0_{10} + 7_{10} = 7_{10})$.

Representing positive numbers in this scheme (with the fourth bit designated as the negative weight) is no different from that of ordinary binary notation.

However, representing negative numbers is not quite as straightforward:

Zero	0000	Negative one	1111
Positive one	0001	Negative two	1110
Positive two	0010	Negative three	1101
Positive three	0011	Negative four	1100
Positive four	0100	Negative five	1011
Positive five	0101	Negative six	1010
Positive six	0110	Negative seven	1001
Positive seven	0111	Negative eight	1000

Note that the negative binary numbers in the right column, being the sum of the right three bits' total plus the negative eight of the leftmost bit, don't "count" in the same progression as the positive binary numbers in the left column.

Rather, the right three bits have to be set at the proper value to equal the desired (negative) total when summed with the negative eight place value of the leftmost bit. Those right three bits are referred to as the *two's complement* of the corresponding positive number.

Consider the following comparison:

Positive Number	Two's Complement
001	111
010	110
011	101

100	100
101	011
110	010
111	001

In this case, with the negative weight bit being the fourth bit (place value of negative eight), the two's complement for any positive number will be wthatever value is needed to add to negative eight to make that positive value's negative equivalent.

Thankfully, there's an easy way to figure out the two's complement for any binary number: simply invert all the bits of that number, changing all 1's to 0's and visa-versa (to arrive at what is called the *one's complement*) and then add one! For example, to obtain the two's complement of five (101_2), we would first invert all the bits to obtain 010_2 (the "one's complement"), then add one to obtain 011_2, or -5_{10} in three-bit, two's complement form.

Interestingly enough, generating the two's complement of a binary number works the same if you manipulate *all* the bits, including the leftmost (sign) bit at the same time as the magnitude bits.

Let's try this with the former example, converting a positive five to a negative five, but performing the complementation process on all four bits.

We must be sure to include the 0 (positive) sign bit on the original number, five (0101_2). First, inverting all bits to obtain the one's complement: 1010_2. Then, adding one, we obtain the final answer: 1011_2, or -5_{10} expressed in four-bit, two's complement form.

It is critically important to remember that the place of the negative-weight bit must be already determined before any two's complement conversions can be done.

If our binary numeration field were such that the eighth bit was designated as the negative-weight bit (10000000_2), we'd have to determine the two's complement based on all seven of the other bits. Here, the two's complement of five (0000101_2) would be 1111011_2. A positive five in this system would be represented as 00000101_2, and a negative five as 11111011_2.

SUBTRACTION

We can subtract one binary number from another by using the standard techniques adapted for decimal numbers (subtraction of each bit pair, right to left, "borrowing" as needed from bits to the left).

However, if we can leverage the already familiar (and easier) technique of binary addition to subtract, that would be better. As we just learned, we can represent negative binary numbers by using the "two's complement" method and a negative place-weight bit. Here, we'll use those negative binary numbers to subtract through addition.

Here's a sample problem:

Subtraction: $7_{10} - 5_{10}$ *Addition equivalent*: $7_{10} + (-5_{10})$

If all we need to do is represent seven and negative five in binary (two's complemented) form, all we need is three bits plus the negative-weight bit:

Positive seven = 0111_2

Negative five = 1011_2

Now, let's add them together:

```
          111  <– Carry bits
          0111
    +     1011
    -----
         10010
    |
Discard extra bit.
```

Answer = 0010_2

Since we've already defined our number bit field as three bits plus the negative-weight bit, the fifth bit in the answer (1) will be discarded to give us a result of 0010_2, or positive two, which is the correct answer.

Another way to understand why we discard that extra bit is to remember that the leftmost bit of the lower number possesses a negative weight, in this case equal to negative eight. When we add these two binary numbers together, what we're actually doing with the MSBs is subtracting the lower number's MSB from the upper number's MSB. In subtraction, one never "carries" a digit or bit on to the next left place-weight.

Let's try another example, this time with larger numbers. If we want to add -25_{10} to 18_{10}, we must first decide how large our binary bit field must be.

To represent the largest (absolute value) number in our problem, which is twenty-five, we need at least five bits, plus a sixth bit for the negative-weight bit. Let's start by representing positive twenty-five, then finding the two's complement and putting it all together into one numeration:

$+25_{10} = 011001_2$ (showing all six bits)

One's complement of $11001_2 = 100110_2$

One's complement + 1 = two's complement = 100111_2

$-25_{10} = 100111_2$

Essentially, we're representing negative twenty-five by using the negative-weight (sixth) bit with a value of negative thirty-two, plus positive seven (binary 111_2).

Now, let's represent positive eighteen in binary form, showing all six bits:

$18_{10} = 010010_2$.

Now, let's add them together and see what we get:.

11 <— Carry bits

100111

+ 010010

————

111001

Since there were no "extra" bits on the left, there are no bits to discard. The leftmost bit on the answer is a 1, which means that the answer is negative, in two's complement form, as it should be. Converting the answer to decimal form by summing all the bits times their respective weight values, we get:

$(1 \times -32_{10}) + (1 \times 16_{10}) + (1 \times 8_{10}) + (1 \times 1_{10}) = -7_{10}$

Indeed -7_{10} is the proper sum of -25_{10} and 18_{10}.

MINIMIZATION OF BOOLEAN FUNCTIONS

The most obvious way to simplify Boolean expressions is

to manipulate them in the same way as normal algebraic expressions are manipulated. With regards to logic relations in digital forms, a set of rules for symbolic manipulation is needed in order to solve for the unknowns.

A set of rules formulated by the English mathematician George Boole describe certain propositions whose outcome would be either true or false. With regard to digital logic, these rules are used to describe circuits whose state can be either, 1 (true) or 0 (false). In order to fully understand this, the relation between the AND gate, OR gate and NOT gate operations should be appreciated. A number of rules can be derived from these relations as demonstrates.

Boolean Postulates:

- *P1*: X = 0 or X = 1
- *P2*: 0. 0 = 0
- *P3*: 1 + 1 = 1
- *P4*: 0 + 0 = 0
- *P5*: 1. 1 = 1
- *P6*: 1. 0 = 0. 1 = 0
- *P7*: 1 + 0 = 0 + 1 = 1

LAWS OF BOOLEAN ALGEBRA

Table shows the basic Boolean laws. Note that every law has two expressions, (a) and (b). This is known as *duality*. These are obtained by changing every AND(.) to OR(+), every OR(+) to AND(.) and all 1's to 0's and vice-versa. It has become conventional to drop the . (AND symbol) i.e. A.B is written as AB.

Boolean Laws

T1: Commutative Law

(a) $A + B = B + A$

(b) $A\,B = B\,A$

T2: Associate Law

(a) $(A + B) + C = A + (B + C)$

(b) $(A\,B)\,C = A\,(B\,C)$

T3: Distributive Law

(a) $A\,(B + C) = A\,B + A\,C$

(b) $A + (B\ C) = (A + B)\ (A + C)$

T4: Identity Law

(a) $A + A = A$

(b) $A\ A = A$

T5: (a) $AB + A\overline{B} = A$

(b) $(A + B)(A + \overline{B}) = A$

T6: Redundance Law

(a) $A + A\ B = A$

(b) $A\ (A + B) = A$

T7: (a) $0 + A = A$

(b) $0\ A = 0$

T8: (a) $1 + A = 1$

(b) $1\ A = A$

T9: (a) $\overline{A} + A = 1$

(b) $\overline{A}\,A = 0$

T10:(a) $A + \overline{A}B = A + B$

(b) $A(\overline{A} + B) = AB$

T11: De Morgan's Theorem

(a) $(\overline{A + B}) = \overline{A}\,\overline{B}$

(b) $(\overline{AB}) = \overline{A} + \overline{B}$

Examples

Prove T10: (a) $A + \overline{A}B = A + B$

- *Algebraically:*

$$\begin{aligned} A + \overline{A}B &= A1 + \overline{A}B && \text{T7(a)} \\ &= A(1 + B) + \overline{A}B && \text{T7(c)} \\ &= A + AB + \overline{A}B && \text{T3(a)} \\ &= A + B\left(A + \overline{A}\right) && \text{T3(a)} \\ &= A + B && \text{T(8)} \end{aligned}$$

- *Using the truth table:*

A	**B**	**A+B**	$\overline{A}B$	$A + \overline{A}B$
0	0	0	0	0
0	1	1	1	1
1	0	1	0	1
1	1	1	0	1

Using the laws given above, complicated expressions can be simplified.

$$Z = \left(A + \overline{B} + \overline{C}\right)\left(A + \overline{B}C\right)$$

$$Z = AA + A\overline{B}C + A\overline{B} + \overline{B}\overline{B}C + A\overline{C} + \overline{B}C\overline{C}$$

$$Z = A\left(1 + \overline{B}C + \overline{B} + \overline{C}\right) + \overline{B}C + \overline{B}C\overline{C} \text{ From laws TBb and Tb}$$

$$Z = A + \overline{B}C \text{ From laws T8b and T9b}$$

KARNAUGH MAPS

So far we can see that applying Boolean algebra can be awkward in order to simplify expressions. Apart from being laborious (and requiring the remembering all the laws) the method can lead to solutions which, though they appear minimal, are not. The Karnaugh map provides a simple and straight-forward method of minimising boolean expressions. With the Karnaugh map Boolean expressions having up to four and even six variables can be simplified. .

The Karnaugh map (K map) provides a systematic method for simplifying a Boolean expression or a truth table function. When used properly, the K map will produce the simplest SOP or POS expression possible. Familiarity with the law and rules of Boolean algebra is not required. Instead, simplification is done graphically using the K mapping technique.

The K map is a table consisting of $N = 2^n$ cells, where n is the number of input variables. For a SOP expression each cell represents one particular combination of the variables in product form. The table format is such that there is a single variable change between any adjacent cells.

This is the characteristic the will determine adjacency. To illustrate the above points let us consider an example where n = 2 and N = 4. Assuming the input variable are A and B then the K map illustrating the four (4) possible variable combinations $\overline{A}\overline{B}, \overline{A}B, A\overline{B}$ and AB is show below in Table.

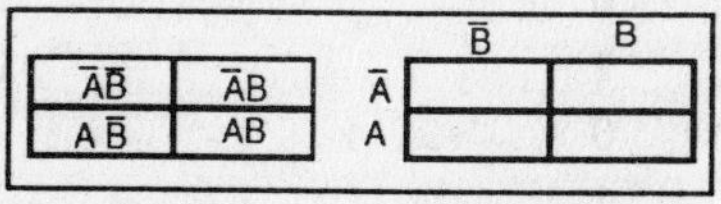

Fig. Two Variable K Map Format.

If we extend our example to consider the case where n = 3 and N = 8 then assuming that our input variables are A, B and C the associated K Map is shown in Table.

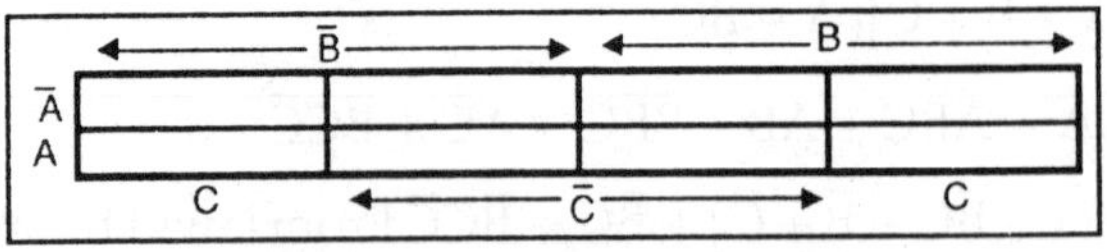

Fig. Three Variable K-Map.

For n>5 the K map technique becomes impractical unless implemented on computer.

Simplifying Using the K-Map

To simplify a SOP for of a Boolean expression using a K map, first identify all the input combinations that produce an output of logic level 1 and place them in their appropriate K map cell. Consequently, all other cells must contain zero (0). Second, group the adjacent cells that contain 1 in a manner that maximises the size of the groups but also minimises the total number of groups. All 1's in the output must be included in a group even if the group is only one cell. Third, as each SOP term represents an AND expression, each (AND) grouping is written with only the input variables that are common to the group. Finally, the simplified expression is formed by ORing each of the (AND) groups. To illustrate lets consider the function, $X = \overline{A}BC + AB\overline{C} + ABC$ whose truth table and K map are illustrated in Tables.

Truth Table of $X = \overline{A}BC + AB\overline{C} + ABC$

Input			Output
A	**B**	**C**	**X**
0	0	0	0
0	0	1	0
0	1	0	0
0	1	1	1
1	0	0	0
1	0	1	0
1	1	0	1
1	1	1	1

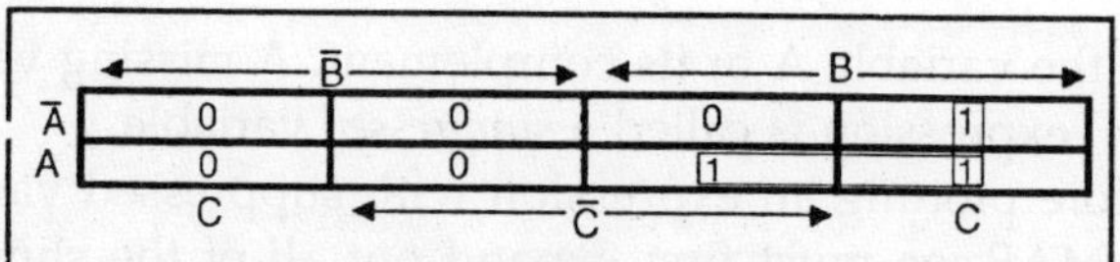

K-Map for $X=\bar{A}BC+AB\bar{C}+ABC$

By examination of the K map, the simplified expression for $X=\bar{A}BC+AB\bar{C}+ABC$ is $X=BC+AB$

Let us now use the K map technique to simplify the SOP Boolean expression $X=\bar{A}\bar{B}\bar{C}+A\bar{B}\bar{C}+\bar{A}\bar{B}C+A\bar{B}C+ABC$. The associated truth table and K map are presented in Table 1-19 and Table, respectively.

Table. Truth $X=\bar{A}\bar{B}\bar{C}+A\bar{B}\bar{C}+\bar{A}\bar{B}C+A\bar{B}C+ABC$

Input			**Output**
A	**B**	**C**	**X**
0	0	0	1
0	0	1	1
0	1	0	0
0	1	1	0
1	0	0	1
1	0	1	1
1	1	0	0
1	1	1	1

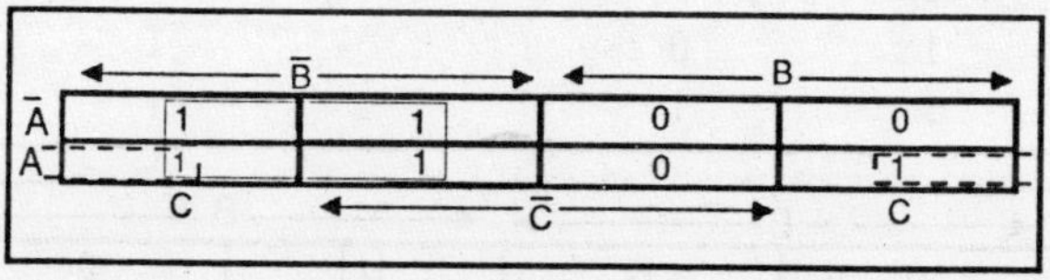

K-Map for $X=\bar{A}\bar{B}\bar{C}+A\bar{B}\bar{C}+\bar{A}\bar{B}C+A\bar{B}C+ABC$

By examination of the K map, the simplified expression for $X=\bar{A}\bar{B}\bar{C}+A\bar{B}\bar{C}+\bar{A}\bar{B}C+A\bar{B}C+ABC$ is $X=\bar{B}+AC$.

Suppressed Variables

Sometimes a SOP expression does not have a complete set of variables in each of its terms. For example, the expression $X=A\bar{B}\bar{C}+B\bar{C}+ABC$ contains a term that does not include the variable A or its complement. A missing variable in a SOP expression is called a *suppressed* variable.

Before plotting an expression with suppressed variables on a K-MAP we must first expand out all of the shortened terms to include the missing variables and its complement.

To illustrate let us consider use our example above:

$$B\bar{C} = B\bar{C}(1)$$

$$B\bar{C} = B\bar{C}\left(A + \bar{A}\right)$$

$$B\bar{C} = AB\bar{C} + \bar{A}B\bar{C}$$

Then our expanded expression becomes,

$$X = A\bar{B}\bar{C} + AB\bar{C} + \bar{A}B\bar{C} + ABC$$

and we may now use the truth table and the associated K map, as illu,strated in Tables to simplify.

Truth Table of $X = A\bar{B}\bar{C} + AB\bar{C} + \bar{A}B\bar{C} + ABC$

Input A	Input B	Input C	Output X
0	0	0	0
0	0	1	0
0	1	0	1
0	1	1	0
1	0	0	1
1	0	1	0
1	1	0	1
1	1	1	1

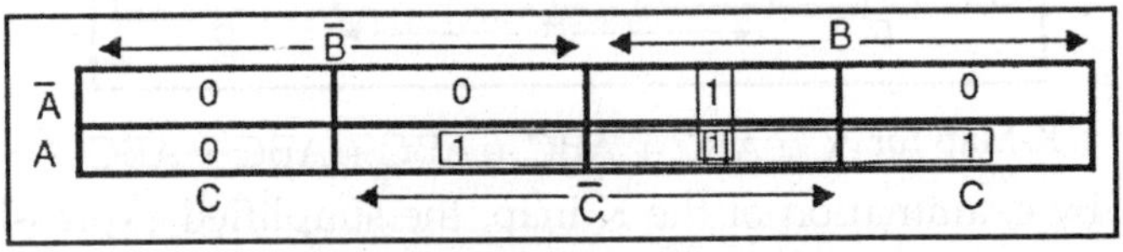

K-Map for $X = A\bar{B}\bar{C} + AB\bar{C} + \bar{A}B\bar{C} + ABC$

From the three groupings shown in the K map we form the expression $X = B\bar{C} + A\bar{C} + AB$.

DON'T CARE STATES

Truth table specifications for a logic function may not to

include all possible combinations of the input binary digits for the input variables, yet they may still be complete specifications of the logic function for the prescribed application. In these situations certain input combinations will not occur due to the nature of the application. When the input combinations are irrelevant or cannot occur, the output states are in the Truth table and the K map are filled with an X and are referred to as don't care states.

When simplifying K maps with don't care states, the contents of the undefined cells (1 or 0) are chosen according to preference. The aim is to enlarge group sizes thereby eliminating as many input variables from the simplified expression as possible. Only those X's that assist in simplifying the function should be included in the groupings. No additional X's should be added that would result in additional terms in the expression.

To illustrate let us consider the function specified by Table 1-23 and its corresponding K map shown in Table. Note that the two groupings determine that the simplified expression is expressed as $J = C + \overline{A}\overline{B}$

Truth Table of the Function J.

Input			Output
A	**B**	**C**	**X**
0	0	0	1
0	0	1	1
0	1	0	0
0	1	1	1
1	0	0	0
1	0	1	X
1	1	0	X
1	1	1	X

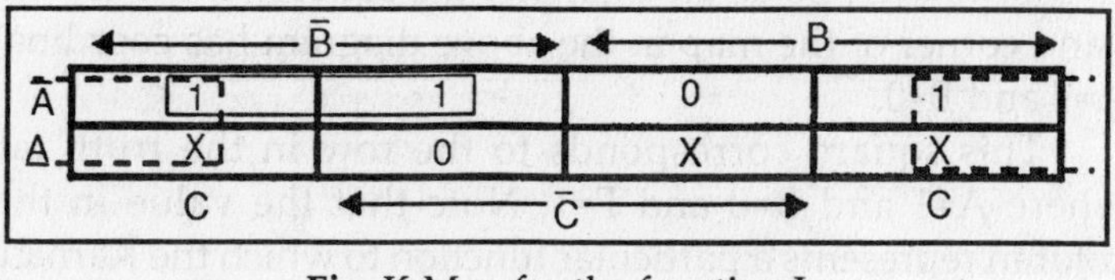

Fig. K-Map for the function J

A Karnaugh map provides a pictorial method of grouping together expressions with common factors and therefore eliminating unwanted variables.

The Karnaugh map can also be described as a special arrangement of a truth table. The diagram below illustrates the correspondence between the Karnaugh map and the truth table for the general case of a two variable problem.

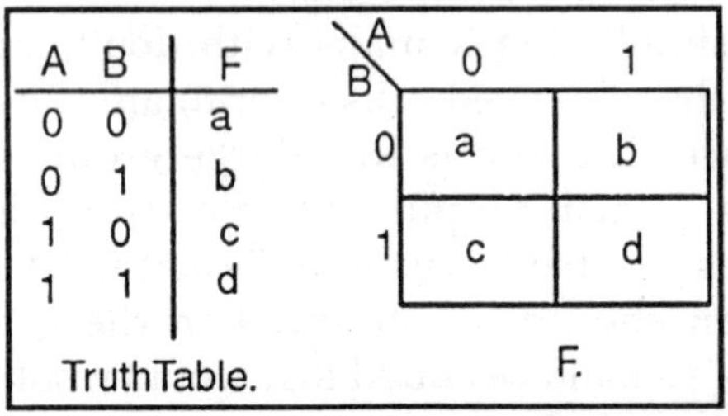

The values inside the squares are copied from the output column of the truth table, therefore there is one square in the map for every row in the truth table.

Around the edge of the Karnaugh map are the values of the two input variable. A is along the top and B is down the left hand side.

The diagram below explains this:

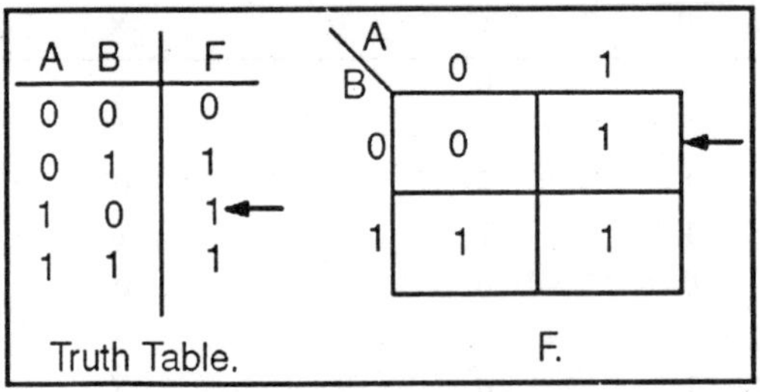

The values around the edge of the map can be thought of as coordinates. So as an example, the square on the top right hand corner of the map in the above diagram has coordinates A=1 and B=0.

This square corresponds to the row in the truth table where A=1 and B=0 and F=1. Note that the value in the F column represents a particular function to which the Karnaugh map corresponds.

EXAMPLES

Example 1:

Consider the following map.

The function plotted is:

$$Z = f(A, B) = A\overline{B} + AB$$

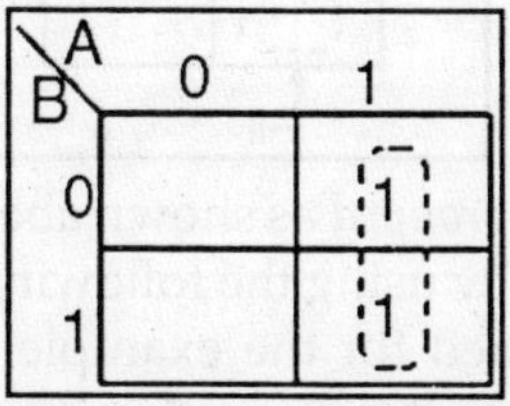

- Note that values of the input variables form the rows and columns. That is the logic values of the variables A and B (with one denoting true form and zero denoting false form) form the head of the rows and columns respectively.
- Bear in mind that the above map is a one dimensional type which can be used to simplify an expression in two variables.
- There is a two-dimensional map that can be used for up to four variables, and a three-dimensional map for up to six variables.

Using algebraic simplification:

$$Z = A\overline{B} + AB$$

$$Z = A(\overline{B} + B)$$

$$Z = A$$

Referring to the map above, the two adjacent 1's are grouped together. Through inspection it can be seen that variable B has its true and false form within the group.

This eliminates variable B leaving only variable A which only has its true form. The minimised answer therefore is Z = A.

Example 2:

Consider the expression $Z = f(A, B) = \overline{A}\overline{B} + A\overline{B} + \overline{A}B$ *plotted on the Karnaugh map:*

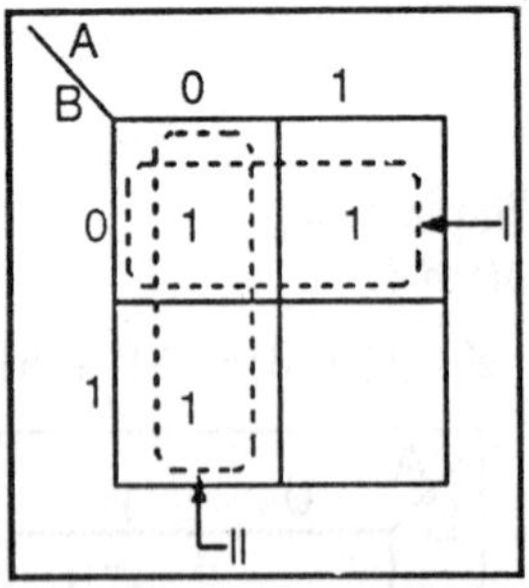

Pairs of 1's are *grouped* as shown above, and the simplified answer is obtained by using the following steps: Note that two groups can be formed for the example given above, bearing in mind that the largest rectangular clusters that can be made consist of two 1s. Notice that a 1 can belong to more than one group. The first group labelled I, consists of two 1s which correspond to A = 0, B = 0 and A = 1, B = 0. Put in another way, all squares in this example that correspond to the area of the map where B = 0 contains 1s, independent of the value of A. So when B = 0 the output is 1. The expression of the output will contain the term $\overline{B}$. For group labelled II corresponds to the area of the map where A = 0. The group can therefore be defined as $\overline{A}$. This implies that when A = 0 the output is 1. The output is therefore 1 whenever B = 0 and A = 0

Hence the simplified answer is $Z = \overline{A} + \overline{B}$

Problems

Minimise the following problems using the Karnaugh maps method:

$$Z = f(A,B,C) = \overline{A}\overline{B}\overline{C} + \overline{A}B + AB\overline{C} + AC$$

$$Z = f(A,B,C) = \overline{A}B + B\overline{C} + BC + A\overline{B}\overline{C}$$

$$Z = f(A,B,C) = \overline{A}\overline{B}\overline{C} + \overline{A}B + AB\overline{C} + AC$$

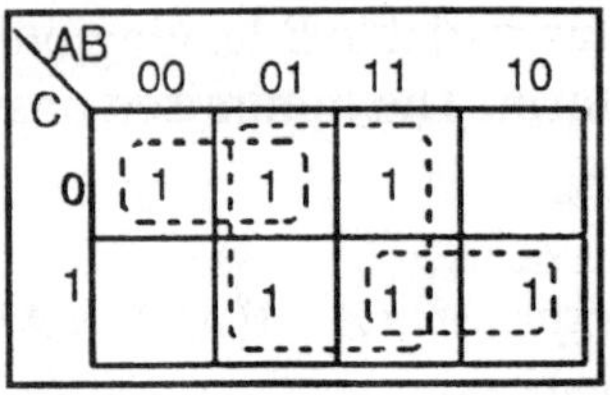

By using the rules of simplification and ringing of adjacent cells in order to make as many variables redundant, the minimised result obtained is:

$B + AC + \overline{A}\overline{C}$ $\quad Z = f(A,B,C) = \overline{A}B + B\overline{C} + BC + A\overline{B}\overline{C}$

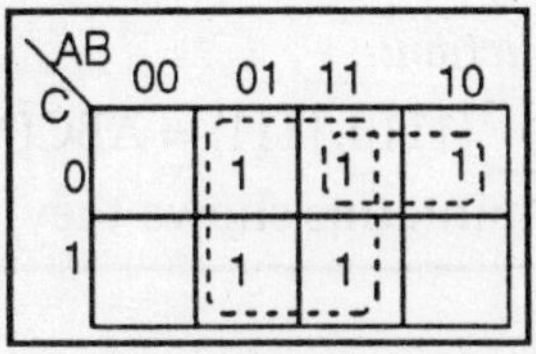

By using the rules of simplification and ringing of adjacent cells in order to make as many variables redundant, the minimised result obtained is $B + A\overline{C}$

TABULAR METHOD OF MINIMISATION

In order to understand the tabular method of minimisation, it is best you understand the numerical assignment of Karnaugh map cells and the incompletely specified functions also known as the can't happen conditions. This is because the tabular method is based on these principles. The tabular method which is also known as the Quine-McCluskey method is particularly useful when minimising functions having a large number of variables, e.g. The six-variable functions.

Computer programmes have been developed employing this algorithm. The method reduces a function in standard sum of products form to a set of prime implicants from which as many variables are eliminated as possible.

These prime implicants are then examined to see if some are redundant. The tabular method makes repeated use of the law $A + \overline{A} = 1$.

Note that Binary notation is used for the function, although decimal notation is also used for the functions. As usual a variable in true form is denoted by 1, in inverted form by 0, and the absence of a variable by a dash (–).

Rules of Tabular Method

Consider a function of three variables f(A, B, C):

$\bar{A}BC$ is represented by 011 ← Binary notation, where A = 0, B = 1 and C = 1

$A\bar{B}\bar{C}$ is represented by 100

$A\bar{C}$ is represented by 1 – 0

BC is represented by – 11

Consider the function:

$$F(A,B,C,D) = \sum(1110, 1111) = ABC\bar{D} + ABCD = ABC$$

Listing the two minterms shows they can be combined,

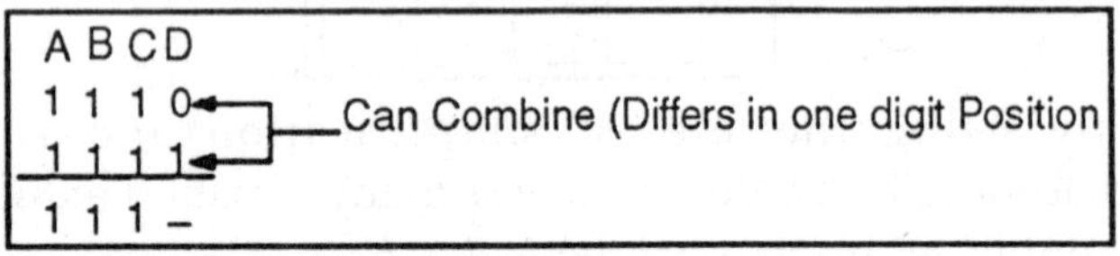

Now consider the following:

$$F(A,B,C,D) = \sum(1101, 1110) = AB\bar{C}D + ABC\bar{D}$$

Note that these variables cannot be combined,

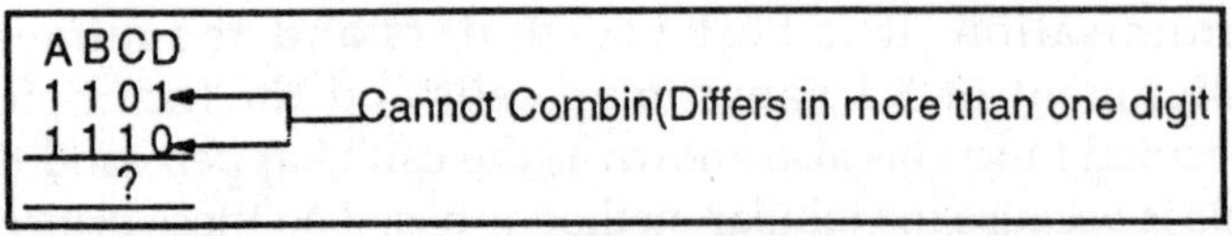

This is because the *FIRST RULE* of the Tabular method for two terms to combine, and thus eliminate one variable, is that they must differ in only one digit position.

Bear in mind that when two terms are combined, one of the combined terms has one digit more at logic 1 than the other combined term. This indicates that the number of 1's in a term is significant and is referred to as its index.

For example: f(A, B, C, D)

0000....................Index 0

0010, 1000.............Index 1

1010, 0011, 1001.......Index 2

1110, 1011.............Index 3

1111...................Index 4

The necessary condition for combining two terms is that the indices of the two terms must differ by one logic variable which must also be the same.

Examples 1:

Consider the function:

$$Z = f(A,B,C) = \overline{A}\overline{B}\overline{C} + \overline{A}\overline{B}C + A\overline{B}\overline{C} + A\overline{B}C$$

To make things easier, change the function into binary notation with index value and decimal value.

$$f(A,B,C) = \sum(000,001,100,101) \text{ — Binary notation}$$

$$0 \quad 1 \quad 1 \quad 2 \text{ — Index}$$

$$0 \quad 1 \quad 4 \quad 5 \text{ — Decimal value}$$

Tabulate the index groups in a colunm and insert the decimal value alongside.

	First List (A B C)	Second List (A B C)	Third List (A B C)
Index 0 → 0	0 0 0 ✓	0,1 0 0 – ✓	0,1,4,5 – 0 –
Index 1 → 1	0 0 1 ✓	0,4 – 0 0 ✓	0,4,1,5 – 0 –
Index 1 → 4	1 0 0 ✓	1,5 – 0 1 ✓	
Index 2 → 5	1 0 0 ✓	4,5 1 0 – ✓	

From the first list, we combine terms that differ by 1 digit only from one index group to the next. These terms from the first list are then seperated into groups in the second list. Note that the ticks are just there to show that one term has been combined with another term. From the second list we can see that the expression is now reduced to:

$$Z = \overline{A}\overline{B} + \overline{B}\overline{C} + \overline{B}C + A\overline{B}$$

From the second list note that the term having an index of 0 can be combined with the terms of index 1. Bear in mind that the dash indicates a missing variable andmust line up in order to get a third list.

The final simplified expression is:

$$Z = \overline{B}$$

Bear in mind that any unticked terms in any list must be included in the final expression (none occured here except from the last list). Note that the only prime implicant here is $Z = \overline{B}$. The tabular method reduces the function to a set of prime implicants. Note that the above solution can be derived algebracially. Attempt this in your notes.

Example 2:

Consider the function:

f(A, B, C, D) = $\sum$(0,1,2,3,5,7,8,10,12,13,15), note that this is in decimal form.

$\sum$(0000,0001,0010,0011,0101,0111,1000,1010,1100,1101,1111) in binary form. (0,1,1,2,2,3,1,2,2,3,4) in the index form.

First List

	A B C D	
0	0 0 0 0	✓
1	0 0 0 1	✓
2	0 0 1 0	✓
8	1 0 0 0	✓
3	0 0 1 1	✓
5	0 1 0 1	✓
10	1 0 1 0	✓
12	1 1 0 0	✓
7	0 1 1 1	✓
13	1 1 1 1	✓
15	1 1 1 1	✓

Second List

	A B C D	
0,1	0 0 0 –	✓
0,2	0 0 – 0	✓
0,8	– 0 0 0	✓
1,3	0 0 – 1	✓
1,5	0 – 0 1	✓
2,3	0 0 1 –	✓
2,10	– 0 1 0	✓
8,10	1 0 – 0	✓
8,12	1 – 0 0	$A\bar{C}\bar{D}$
3,7	0 – 1 1	✓
5,7	0 1 – 1	✓
5,13	– 1 0 1	✓
12,13	1 1 0 –	$AB\bar{C}$
7,15	– 1 1 1	✓
13,15	1 1 – 1	✓

Third List

	A B C D	
0,1 2 3	0 0 – –	$\bar{A}\bar{B}$
0,2 1 3	0 0 – –	
0,2,8,10	– 0 – 0	$\bar{B}\bar{D}$
0,8,2,10	– 0 0	
1,3,5,7	0 – – 1	$\bar{A}D$
1,5,3,7	0 – – 1	
5,7,13,15	– 1 – 1	BD
5,13,7,15	– 1 – 1	

The prime implicants are:

$$\bar{A}\bar{B}+\bar{B}\bar{D}+\bar{A}D+BD+A\bar{C}\bar{D}+AB\bar{C}$$

The chart is used to remove redundant prime implicants. A grid is prepared having all the prime implicants listed at the left and all the minterms of the function along the top. Each minterm covered by a given prime implicant is marked in the appropriate position.

	0	1	2	3	5	7	8	10	12	13	15
$\bar{A}\bar{B}$	X	X	X	X							
$\bar{B}\bar{D}$	X		X				X	⊗			
$\bar{A}D$		X		X	X	X					
BD					X	X				X	⊗
$A\bar{C}\bar{D}$							X		X		
$AB\bar{C}$									X	X	
Bssential	X		X		X	X	X	⊗		X	⊗

From the above chart, BD is an essential prime implicant. It is the only prime implicant that covers the minterm decimal 15 and it also includes 5, 7 and 13 $\overline{B}\overline{D}$ is also an essential prime implicant. It is the only prime implicant that covers the minterm denoted by decimal 10 and it also includes the terms 0, 2 and 8.

The other minterms of the function are 1, 3 and 12. Minterm 1 is present in $\overline{A}\overline{B}$ and $\overline{A}D$. Similarly for minterm 3. We can therefore use either of these prime implicants for these minterms. Minterm 12 is present in $A\overline{C}\overline{D}$ and $AB\overline{C}$, so again either can be used.

Thus, one minimal solution is:

$$Z = \overline{B}\overline{D} + BD + \overline{A}\overline{B} + A\overline{C}\overline{D}$$

Problems

1. *Minimise the function below using the tabular method of simplification:*

$$Z = f(A,B,C,D)\ \overline{A}\overline{B}\overline{C}\overline{D} + \overline{A}\overline{B}C\overline{D}$$

$$+ A\overline{B}CD + A\overline{B}C\overline{D} + \overline{A}BCD + \overline{A}BC\overline{D} + \overline{A}\overline{B}CD$$

2. Using the tabular method of simplification, find all equally minimal solutions for the function below.
$Z = f(A,B,C,D) = \sum(1,4,5,10,12,14)$
3. *Consider the function:*
$Z = f(A,B,C,D)\ \overline{A}\overline{B}\overline{C}\overline{D} + \overline{A}\overline{B}C\overline{D} +$

$A\overline{B}CD + A\overline{B}C\overline{D} + \overline{A}BCD + \overline{A}BC\overline{D} + \overline{A}\overline{B}CD$

Convert to decimal and binary equivalents:

$Z = f(A,B,C,D) = \sum(0,2,4,5,8,9,12)$ - decimal

equivalent $Z = f(A,B,C,D)$

$= \sum(0000,0010,0100,0101,1000,1001,1100)$ - binary

equivalent (0, 1, 1, 2, 1, 2, 2) - index values

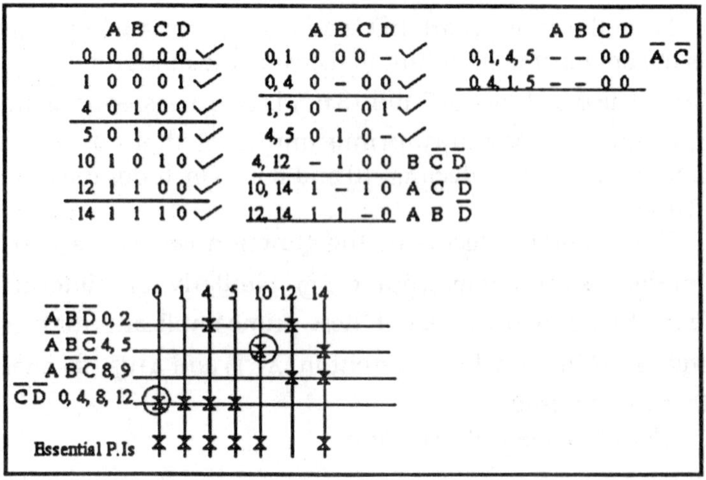

The simplified answer is:

$$Z = Z = \bar{A}\bar{B}\bar{D} + \bar{A}B\bar{C} + A\bar{B}\bar{C} + \bar{C}\bar{D}$$

4. F(A,B,C)=

$$\sum(0,1,4,5,10,12,14,) =$$

$$\sum(0000,0001,0101,1010,1100,1110)$$

$$= \text{Index}: 012223$$

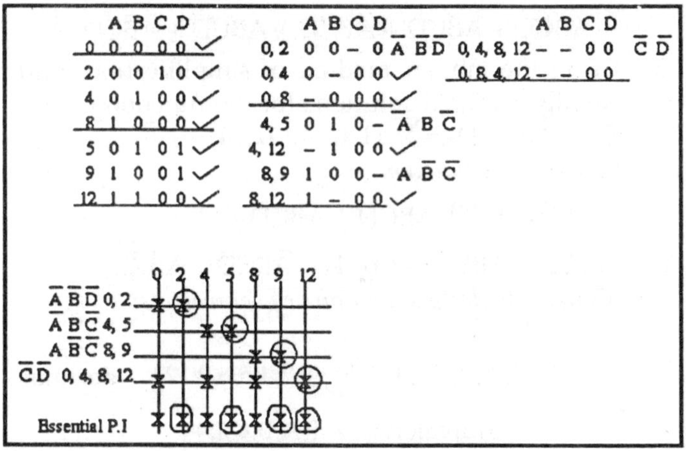

Note that the remaining term 12 is covered by $B\bar{C}\bar{D}$ and $AB\bar{D}$

One simplified answer is:

$$Z = \bar{A}\bar{C} + AC\bar{D} + B\bar{C}\bar{D}$$

Another answer is:

$$Z = \bar{A}\bar{C} + AC\bar{D} + AB\bar{D}$$

CODE CONVERSIONS

Converting from one code form to another code form is called code conversion, like converting from binary to decimal or converting from hexadecimal to decimal.

BINARY-TO-DECIMAL CONVERSION

Any binary number can be converted to its decimal equivalent simply by summing together the weights of the various positions in the binary number which contain a 1.

Binary	Decimal
11011_2	
$2^4+2^3+0^1+2^1+2^0$	=16+8+0+2+1
Result	27_{10}

and

Binary	Decimal
10110101_2	
$2^7+0^6+2^5+2^4+0^3+2^2+0^1+2^0$	
=128+0+32+16+0+4+0+1	
Result	181_{10}

You should have noticed that the method is to find the weights (i.e., powers of 2) for each bit position that contains a 1, and then to add them up.

DECIMAL-TO-BINARY CONVERSION

There are 2 Methods

- Reverse of Binary-To-Decimal Method
- Repeat Division

Reverse of Binary-To-Decimal Method

Decimal	Binary
45_{10}	=32 + 0 + 8 + 4 +0 + 1

	$=2^5+0+2^3+2^2+0+2^0$
Result	$=101101_2$

Repeat Division-Convert Decimal to Binary

This method uses repeated division by 2.
Convert 25_{10} to binary.

Division	Remainder	Binary
25/2	= 12+ remainder of 1	1 (Least Significant Bit)
12/2	= 6 + remainder of 0	0
6/2	= 3 + remainder of 0	0
3/2	= 1 + remainder of 1	1
1/2	= 0 + remainder of 1	1 (Most Significant Bit)
Result	25_{10}	$= 11001_2$

The Flow chart for repeated-division method is as follows:

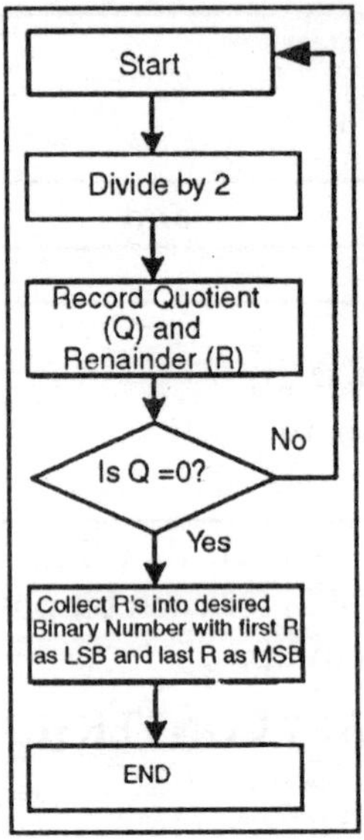

BINARY-TO-OCTAL/OCTAL-TO-BINARY CONVERSION

Octal Digit	0	1	2	3	4	5	6	7
Binary Equivalent	000	001	010	011	100	101	110	111

Each Octal digit is represented by three binary digits.
Examplc:
$100\ 111\ 010_2 = (100)\ (111)\ (010)_2 = 4\ 7\ 2_8$

Repeat Division-Convert Decimal to Octal

This method uses repeated division by 8.
Example: Convert 177_{10} to octal and binary

Division	Result	Binary
177/8	= 22+ remainder of 1	1 (Least Significant Bit)
22/8	= 2 + remainder of 6	6
2/8	= 0 + remainder of 2	2 (Most Significant Bit)
Result	177_{10}	= 261_8
Binary		010110001_2

HEXADECIMAL TO DECIMAL/DECIMAL TO HEXADECIMAL CONVERSION

Example:
$2AF_{16} = 2 \times (16^2) + 10 \times (16^1) + 15 \times (16^0) = 687_{10}$

Repeat Division- Convert Decimal to Hexadecimal

This method uses repeated division by 16.
Example: convert 378_{10} to hexadecimal and binary:

Division	Result	Hexadecimal
378/16	= 23+ remainder of 10	A (Least Significant Bit)23
23/16	= 1 + remainder of 7	7
1/16	= 0 + remainder of 1	1 (Most Significant Bit)
Result	378_{10}	= $17A_{16}$
Binary		= $0001\ 0111\ 1010_2$

BINARY-TO-HEXADECIMAL/HEXADECIMAL-TO-BINARY CONVERSION

Hexadecimal Digit	0	1	2	3	4	5	6	7
Binary Equivalent 0111v	0000	0001	0010	0011	0100	0101	0110	

Hexadecimal Digit	8	9	A	B	C	D	E	F
Binary Equivalent	1000	1001	1010	1011	1100	1101	1110	1111

Each Hexadecimal digit is represented by four bits of binary digit.

Example:

$1011\ 0010\ 1111_2 = (1011)\ (0010)\ (1111)_2 = B\ 2\ F_{16}$

OCTAL-TO-HEXADECIMAL HEXADECIMAL-TO-OCTAL CONVERSION

- Convert Octal (Hexadecimal) to Binary first.
- Regroup the binary number by three bits per group starting from LSB if Octal is required.
- Regroup the binary number by four bits per group starting from LSB if Hexadecimal is required.

Example: Convert $5A8_{16}$ to Octal.

Hexadecimal	Binary/Octal
5A816	= 0101 1010 1000 (Binary)
	= 010 110 101 000 (Binary
Result	= 2 6 5 0 (Octal)

DEMORGANS THEOREM

For N variables, DeMorgan's theorems are expressed in the following formulas:

$$\overline{ABC...N} = \bar{A} + \bar{B} + \bar{C} + ... + \bar{N} (1-2)$$

That is, the complement of the product is equivalent to the sum of the complements,

$$\overline{A + B + C + ... + N} = \bar{A}\bar{B}\bar{C}....\bar{N} (1-3)$$

Similarly, the complement of the sum is equivalent to the product of the complements. Applications of DeMorgan's theorems range from deriving the composite of functions to providing alternative design to logic circuits. The objective of using these theorems in circuit design is to minimise the number of ICs required in a logic circuit.

EXAMPLES OF DEMORGAN'S THEOREM

Applying Demorgan's theorem to the expression,

$$\overline{A+B+C+...+N} = \bar{A}\bar{B}\bar{C}....\bar{N}(1-3)$$

we get:

$$\overline{(A+B+C)D} = \overline{(A+B+C} + \bar{D}$$
$$= \bar{A}\bar{B}\bar{C} + \bar{D}$$

Applying Demorgan's theorem to the expression ABC + DEF we get:

$$\overline{ABC+DEF} = (\overline{ABC})(\overline{DEF})$$
$$= (\bar{A}+\bar{B}+\bar{C})(\bar{D}+\bar{E}+\bar{F})$$

First, recall that a binary variable may be either in its true form (*A*) or its complement ($\bar{A}$). Second, recall that for *n*variables, the maximum number of input variable combinations is given by,

$$N = 2^n.$$

Then considering the AND gate, each of the N logic expressions formed is called a *standard product* or *minterm.*

As indicated in Table , binary digits '1' and '0' are taken to represent a given variable (e.g. *A*) and its complemented (e.g. $\bar{A}$), respectively. Also from Table note that each minterm is assigned a symbol (P_j) each where j is the decimal equivalent to the binary number of the minterm designated.

Similarly, if we consider an OR gate, each of the *N* logic expressions formed is called a *standard sum* or *maxterm.* In this case binary digits '1' and '0' are taken to represent a given complemented variable (e.g. $\bar{A}$) and its true form (e.g.*A*), respectively.

As shown in Table, a symbol (S_j) is assigned to each maxterm where j is the decimal equivalent to the binary number of the maxterm designated.

Also observe that each maxterm is the complement of its corresponding minterm, and vice versa.

Table. The Complement of Each Term may be Determined using DeMorgan's Theorem

Input			Minterms		Maxterms	
A	B	C	Terms	Designation	Terms	Designation
0	0	0	$\overline{A}\,\overline{B}\,\overline{C}$	P_0	$A+B+C$	S_0
0	0	1	$\overline{A}\,\overline{B}\,C$	P_1	$A+B+\overline{C}$	S_1
0	1	0	$\overline{A}\,B\,\overline{C}$	P_2	$A+\overline{B}+C$	S_2
0	1	1	$\overline{A}\,B\,C$	P_3	$A+\overline{B}+\overline{C}$	S_3
1	0	0	$A\,\overline{B}\,\overline{C}$	P_4	$\overline{A}+B+C$	S_4
1	0	1	$A\,\overline{B}\,C$	P_5	$\overline{A}+B+\overline{C}$	S_5
1	1	0	$A\,B\,\overline{C}$	P_6	$\overline{A}+\overline{B}+C$	S_6
1	1	1	$A\,B\,C$	P_7	$\overline{A}+\overline{B}+\overline{C}$	S_7

Table above shows the Minterms and Maxterms for three Binary Variables.

SIGNIFICANCE OF MINTERMS AND MAXTERMS

In short, minterms and maxterms may be used to define the two standard forms for logic expressions, namely the sum of products (SOP), or sum of minterms, and the product of sums (POS), or product of maxterms.

These standard forms of expression aid the logic circuit designer by simplifying the derivation of the function to be implemented.

Sum of Products

The SOP expression is the equation of the logic function as read off the truth table to specify the input combinations when the output is a logical 1. To illustrate, let us consider Table. Observe that the output is high for the rows labelled 3, 5 and 6. The SOP expression for this circuit is thus given any of the following:

1. $X = \overline{A}BC + A\overline{B}C + AB\overline{C}$
2. $X = P_3 + P_5 + P_6$ or
3. $X(A,B,C) = \sum(3,5,6)$

Table. Truth of $X = \bar{A}BC + A\bar{B}C + AB\bar{C}$

Row Number	Input A	Output B
C	X	0
0	0	0
0	1	0
0	1	0
2	0	1
0	0	3
0	1	1
1	4	1
0	0	0
5	1	0
1	1	6
1	1	0
1	7	1
1	1	0

Product of Sums

The POS expression is the equation of the logic function as read off the truth table to specify the input combinations when the output is a logical 0.

To illustrate, let us again consider Table. Observe that the output is low for the rows labelled 0, 1, 2, 4 and 7.

The POS expression for this circuit is thus given by any of the following:

1. $X = (A+B+C)(A+B+\bar{C})(A+\bar{B}+C)(\bar{A}+B+C)$
   $(\bar{A}+\bar{B}+\bar{C})$
2. $X = S_0S_1S_2S_4S_7$
3. $X(A,B,C) = \prod(0,1,2,4,7)$

Further Examples

Derive the SOP and POS expressions for the truth tables shown below:

Table. Truth Describing the Output X

Row Number	Input A	Input B	Output X
0	0	0	1
1	0	1	0
2	1	0	1
3	1	1	0

SOP:

$X = P_0 + P_2$ or

$$X = \bar{A}\bar{B} + A\bar{B}$$

POS:

$X = S_1 S_3$ or,

$$X = (A + \bar{B})(\bar{A} + \bar{B})$$

Recall from an earlier lecture that we stated that the NAND and NOR were universal logic gates. Using DeMorgan's theorem and the Rules and laws of Boolean algebra proving this should be an easy task. Figure shows the equivalency between the basic logic gates and their NAND logic circuits counterpart. Similarly, Figure shows the equivalency between the basic logic gates and their NOR logic circuits counterpart.

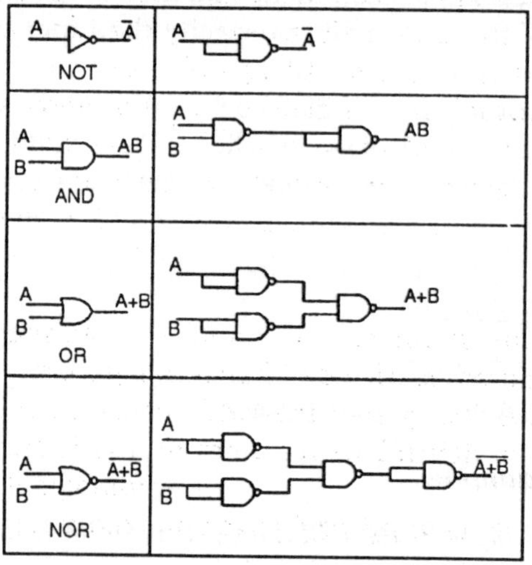

Fig. NAND Equivalent Circuits

Note that $AB = \overline{\overline{AB}}$ i.e the complement of "A NAND B"

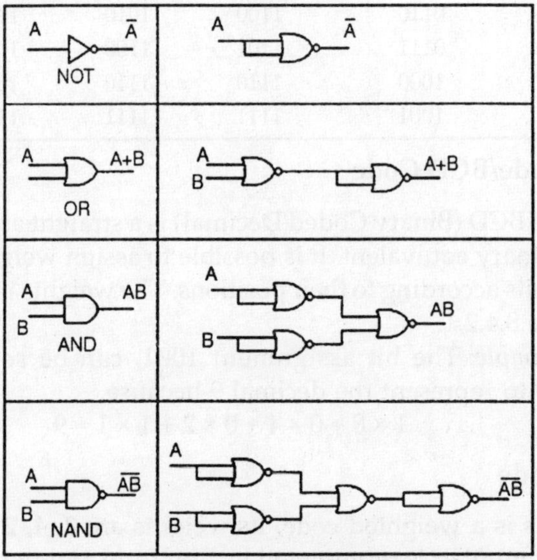

Fig. NOR Equivalent Circuits

BINARY CODES

Binary codes are codes which are represented in binary system with modification from the original ones.

Below we will be seeing the following:

- Weighted Binary Systems
- Non Weighted Codes

WEIGHTED BINARY SYSTEMS

Weighted binary codes are those which obey the positional weighting principles, each position of the number represents a specific weight. The binary counting sequence is an example.

Decimal	8421	2421	5211	Excess-3
0	0000	0000	0000	0011
1	0001	0001	0001	0100
2	0010	0010	0011	0101
3	0011	0011	0101	0110

4	0100	0100	0111	0111
5	0101	1011	1000	1000
6	0110	1100	1010	1001
7	0111	1101	1100	1010
8	1000	1110	1110	1011
9	1001	1111	1111	1100

8421 Code/BCD Code

The BCD (Binary Coded Decimal) is a straight assignment of the binary equivalent. It is possible to assign weights to the binary bits according to their positions. The weights in the BCD code are 8,4,2,1.

Example: The bit assignment 1001, can be seen by its weights to represent the decimal 9 because.

$$1 \times 8 + 0 \times 4 + 0 \times 2 + 1 \times 1 = 9$$

2421 Code

This is a weighted code, its weights are 2, 4, 2 and 1. A decimal number is represented in 4-bit form and the total four bits weight is 2 + 4 + 2 + 1 = 9. Hence the 2421 code represents the decimal numbers from 0 to 9.

5211 Code

This is a weighted code, its weights are 5, 2, 1 and 1. A decimal number is represented in 4-bit form and the total four bits weight is 5 + 2 + 1 + 1 = 9. Hence the 5211 code represents the decimal numbers from 0 to 9.

Reflective Code

A code is said to be reflective when code for 9 is complement for the code for 0, and so is for 8 and 1 codes, 7 and 2, 6 and 3, 5 and 4. Codes 2421, 5211, and excess-3 are reflective, whereas the 8421 code is not.

Sequential Codes

A code is said to be sequential when two subsequent codes, seen as numbers in binary representation, differ by one. This greatly aids mathematical manipulation of data. The 8421 and Excess-3 codes are sequential, whereas the 2421 and 5211 codes are not.

NON WEIGHTED CODES

Non weighted codes are codes that are not positionally weighted. That is, each position within the binary number is not assigned a fixed value.

Excess-3 Code

Excess-3 is a non weighted code used to express decimal numbers. The code derives its name from the fact that each binary code is the corresponding 8421 code plus 0011(3).

Example: 1000 of 8421 = 1011 in Excess-3

Gray Code

The gray code belongs to a class of codes called minimum change codes, in which only one bit in the code changes when moving from one code to the next. The Gray code is non-weighted code, as the position of bit does not contain any weight.

The gray code is a reflective digital code which has the special property that any two subsequent numbers codes differ by only one bit. This is also called a unit-distance code. In digital Gray code has got a special place.

Decimal Number	Binary Code	Gray Code
0	0000	0000
1	0001	0001
2	0010	0011
3	0011	0010
4	0100	0110
5	0101	0111
6	0110	0101
7	0111	0100
8	1000	1100
9	1001	1101
10	1010	1111
11	1011	1110
12	1100	1010

13	1101	1011
14	1110	1001
15	1111	1000

BINARY TO GRAY CONVERSION

- Gray Code MSB is binary code MSB.
- Gray Code MSB-1 is the XOR of binary code MSB and MSB-1.
- MSB-2 bit of gray code is XOR of MSB-1 and MSB-2 bit of binary code.
- MSB-N bit of gray code is XOR of MSB-N-1 and MSB-N bit of binary code.

ERROR DETECTING AND CORRECTING CODES

For reliable transmission and storage of digital data, error detection and correction is required. Below are a few examples of codes which permit error detection and error correction after detection.

ERROR DETECTING CODES

When data is transmitted from one point to another, like in wireless transmission, or it is just stored, like in hard disks and memories, there are chances that data may get corrupted. To detect these data errors, we use special codes, which are error detection codes.

Parity

In parity codes, every data byte, or nibble (according to how user wants to use it) is checked if they have even number of ones or even number of zeros. Based on this information an additional bit is appended to the original data. Thus if we consider 8-bit data, adding the parity bit will make it 9 bit long.

At the receiver side, once again parity is calculated and matched with the received parity (bit 9), and if they match, data is ok, otherwise data is corrupt.

There are two types of parity:

- Even parity: Checks if there is an even number of

ones; if so, parity bit is zero. When the number of ones is odd then parity bit is set to.
- Odd Parity: Checks if there is an odd number of ones; if so, parity bit is zero. When number of ones is even then parity bit is set to.

Check Sums

The parity method is calculated over byte, word or double word. But when errors need to be checked over 128 bytes or more (basically blocks of data), then calculating parity is not the right way. So we have checksum, which allows tc check for errors on block of data. There are many variations of checksum.

- Adding all bytes
- CRC
- Fletcher's checksum
- Adler-32

The simplest form of checksum, which simply adds up the asserted bits in the data, cannot detect a number of types of errors.

In particular, such a checksum is not changed by:

- Reordering of the bytes in the message
- Inserting or deleting zero-valued bytes
- Multiple errors which sum to zero

Example of Checksum: Given 4 bytes of data (can be done with any number of bytes): 25h, 62h, 3Fh, 52h

- Adding all bytes together gives 118h.
- Drop the Carry Nibble to give you 18h.
- Get the two's complement of the 18h to get E8h. This is the checksum byte.

To Test the Checksum byte simply add it to the original group of bytes. This should give you 200h. Drop the carry nibble again giving 00h. Since it is 00h this means the checksum means the bytes were probably not changed.

ERROR-CORRECTING CODES

Error-correcting codes not only detect errors, but also correct them. This is used normally in Satellite communication, where turn-around delay is very high as is the probability of data getting corrupt.

ECC (Error correcting codes) are used also in memories, networking, Hard disk, CDROM, DVD etc. Normally in networking chips (ASIC), we have 2 Error detection bits and 1 Error correction bit.

HAMMING CODE

Hamming code adds a minimum number of bits to the data transmitted in a noisy channel, to be able to correct every possible one-bit error. It can detect (not correct) two-bits errors and cannot distinguish between 1-bit and 2-bits inconsistencies. It can't - in general - detect 3(or more)-bits errors. The idea is that the failed bit position in an n-bit string (which we'll call X) can be represented in binary with $\log_2(n)$ bits, hence we'll try to get it adding just $\log_2(n)$ bits.

First, we set $m = n + \log_2(n)$ to the encoded string length and we number each bit position starting from 1 through m. Then we place these additional bits at power-of-two positions, that is 1, 2, 4, 8..., while remaining ones (3, 5, 6, 7...) hold the bit string in the original order.

Now we set each added bit to the parity of a group of bits. We group bits this way: we form a group for every parity bit, where the following relation holds: position(bit) AND position(parity) = position(parity) (Note that: AND is the bit-wise boolean AND; parity bits are included in the groups; each bit can belong to one or more groups.) So bit 1 groups bits 1, 3, 5, 7... while bit 2 groups bits 2, 3, 6, 7, 10..., bit 4 groups bits 4, 5, 6, 7, 12, 13... and so on. Thus, by definition, X (the failed bit position defined above) is the sum of the incorrect parity bits positions (0 for no errors).

To understand why it is so, let's call X_n the nth bit of X in binary representation. Now consider that each parity bit is tied to a bit of X: parity1 $\rightarrow X_1$, parity2 $\rightarrow X_2$, parity4 $\rightarrow X_3$, parity8 $\rightarrow X_4$ and so on – for programmers: they are the respective AND masks –. By construction, the failed bit makes fail only the parity bits which correspond to the 1s in X, so each bit of X is 1 if the corresponding parity is wrong and 0 if it is correct.

Note that the longer the string, the higher the throughput n/m and the lower the probability that no more than one bit fails. So the string to be sent should be broken into blocks whose length depends on the transmission channel quality (the cleaner the channel, the bigger the block). Also, unless it's guaranteed that at most one bit per block fails, a checksum or some other form of data integrity check should be added.

ALPHANUMERIC CODES

The binary codes that can be used to represent all the letters of the alphabet, numbers and mathematical symbols, punctuation marks, are known as alphanumeric codes or character codes. These codes enable us to interface the input-output devices like the keyboard, printers, video displays with the computer.

ASCII CODE

ASCII stands for American Standard Code for Information Interchange. It has become a world standard alphanumeric code for microcomputers and computers. It is a 7-bit code representing $2^7 = 128$ different characters. These characters represent 26 upper case letters (A to Z), 26 lowercase letters (a to z), 10 numbers (0 to 9), 33 special characters and symbols and 33 control characters.

The 7-bit code is divided into two portions, The leftmost 3 bits portion is called zone bits and the 4-bit portion on the right is called numeric bits. An 8-bit version of ASCII code is known as USACC-II 8 or ASCII-8. The 8-bit version can represent a maximum of 256 characters.

EBCDIC CODE

EBCDIC stands for Extended Binary Coded Decimal Interchange. It is mainly used with large computer systems like mainframes. EBCDIC is an 8-bit code and thus accomodates up to 256 characters. An EBCDIC code is divided into two portions: 4 zone bits (on the left) and 4 numeric bits.

Chapter 2

Concept of Boolean Algebra

OVERVIEW

Mathematical rules are based on the defining limits we place on the particular numerical quantities dealt with. When we say that 1 + 1 = 2 or 3 + 4 = 7, we are implying the use of integer quantities: the same types of numbers we all learned to count in elementary education.

What most people assume to be self-evident rules of arithmetic valid at all times and for all purposes actually depend on what we define a number to be.

For instance, when calculating quantities in AC circuits, we find that the "real" number quantities which served us so well in DC circuit analysis are inadequate for the task of representing AC quantities.

We know that voltages add when connected in series, but we also know that it is possible to connect a 3-volt AC source in series with a 4-volt AC source and end up with 5 volts total voltage (3 + 4 = 5)! Does this mean the inviolable and self-evident rules of arithmetic have been violated? No, it just means that the rules of "real" numbers do not apply to the kinds of quantities encountered in AC circuits, where every variable has both a magnitude and a phase.

Consequently, we must use a different kind of numerical quantity, or object, for AC circuits (*complex* numbers, rather than *real* numbers), and along with this different system of numbers comes a different set of rules telling us how they relate to one another.

An expression such as "3 + 4 = 5" is nonsense within the

scope and definition of real numbers, but it fits nicely within the scope and definition of complex numbers (think of a right triangle with opposite and adjacent sides of 3 and 4, with a hypotenuse of 5).

Because complex numbers are two-dimensional, they are able to "add" with one another trigonometrically as single-dimension "real" numbers cannot.

Logic is much like mathematics in this respect: the so-called "Laws" of logic depend on how we define what a proposition is.

The Greek philosopher Aristotle founded a system of logic based on only two types of propositions: true and false. His bivalent (two-mode) definition of truth led to the four foundational laws of logic: the Law of Identity (A is A); the Law of Non-contradiction (A is not non-A); the Law of the Excluded Middle (either A or non-A); and the Law of Rational Inference.

These so-called Laws function within the scope of logic where a proposition is limited to one of two possible values, but may not apply in cases where propositions can hold values other than "true" or "false." In fact, much work has been done and continues to be done on "multivalued," or *fuzzy* logic, where propositions may be true or false *to a limited degree.*

In such a system of logic, "Laws" such as the Law of the Excluded Middle simply do not apply, because they are founded on the assumption of bivalence. Likewise, many premises which would violate the Law of Non-contradiction in Aristotelian logic have validity in "fuzzy" logic. Again, the defining limits of propositional values determine the Laws describing their functions and relations.

The English mathematician George Boole sought to give symbolic form to Aristotle's system of logic. Boole wrote a treatise on the subject in 1854, titled *An Investigation of the Laws of Thought, on Which Are Founded the Mathematical Theories of Logic and Probabilities,* which codified several rules of relationship between mathematical quantities limited to one of two possible values: true or false, 1 or 0. His mathematical system became known as Boolean algebra.

All arithmetic operations performed with Boolean quantities have but one of two possible outcomes: either 1 or 0. There is no such thing as "2" or "–1" or "1/2" in the Boolean world. It is a world in which all other possibilities are invalid by fiat. As one might guess, this is not the kind of math you want to use when balancing a checkbook or calculating current through a resistor.

However, Claude Shannon of MIT fame recognized how Boolean algebra could be applied to on-and-off circuits, where all signals are characterized as either "high" (1) or "low" (0). The thesis, titled *A Symbolic Analysis of Relay and Switching Circuits*, put Boole's theoretical work to use in a way Boole never could have imagined, giving us a powerful mathematical tool for designing and analyzing digital circuits.

In this chapter, you will find a lot of similarities between Boolean algebra and "normal" algebra, the kind of algebra involving so-called real numbers. Just bear in mind that the system of numbers defining Boolean algebra is severely limited in terms of scope, and that there can only be one of two possible values for any Boolean variable: 1 or 0.

Consequently, the "Laws" of Boolean algebra often differ from the "Laws" of real-number algebra, making possible such statements as 1 + 1 = 1, which would normally be considered absurd. Once you comprehend the premise of all quantities in Boolean algebra being limited to the two possibilities of 1 and 0, and the general philosophical principle of Laws depending on quantitative definitions, the "nonsense" of Boolean algebra disappears.

It should be clearly understood that Boolean numbers are not the same as *binary* numbers. Whereas Boolean numbers represent an entirely different system of mathematics from real numbers, binary is nothing more than an alternative *notation*for real numbers.

The two are often confused because both Boolean math and binary notation use the same two ciphers: 1 and 0. The difference is that Boolean quantities are restricted to a single bit (either 1 or 0), whereas binary numbers may be composed of many bits adding up in place-weighted form to a value of any finite size. The binary number 10011_2("nineteen") has no more place in the Boolean world than the decimal number 2_{10} ("two") or the octal number 32_8 ("twenty-six").

BOOLEAN ARITHMETIC

Let us begin our exploration of Boolean algebra by adding numbers together:

$$0+0=0$$
$$0+1=1$$
$$1+0=1$$
$$1+1=1$$

The first three sums make perfect sense to anyone familiar with elementary addition. The last sum, though, is quite possibly responsible for more confusion than any other single statement in digital electronics, because it seems to run contrary to the basic principles of mathematics. Well, it *does* contradict principles of addition for real numbers, but not for Boolean numbers.

Remember that in the world of Boolean algebra, there are only two possible values for any quantity and for any arithmetic operation: 1 or 0.

There is no such thing as "2" within the scope of Boolean values. Since the sum "1 + 1" certainly isn't 0, it must be 1 by process of elimination.

It does not matter how many or few terms we add together, either. Consider the following sums:

$$0+1+1=1$$
$$1+1+1=1$$
$$0+1+1+1=1$$
$$1+0+1+1+1=1$$

Take a close look at the two-term sums in the first set of equations. Does that pattern look familiar to you? It should! It is the same pattern of 1's and 0's as seen in the truth table for an OR gate. In other words, Boolean addition corresponds to the logical function of an "OR" gate, as well as to parallel switch contacts:

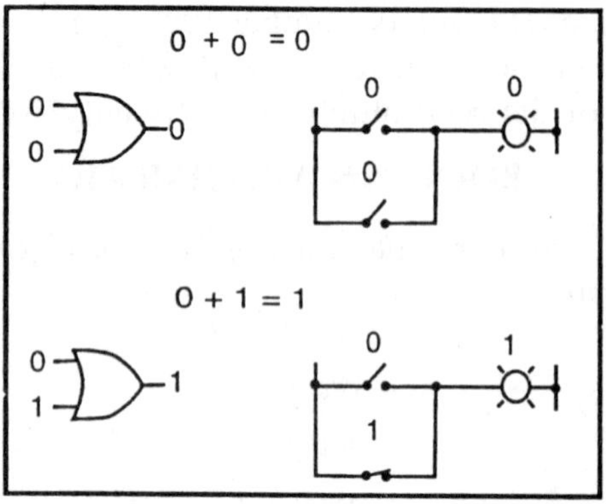

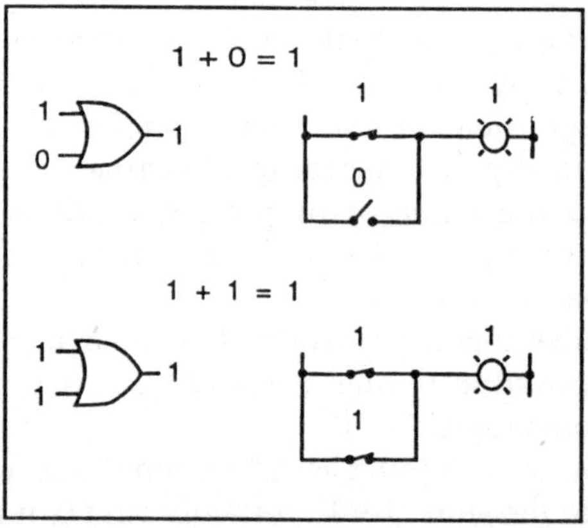

There is no such thing as subtraction in the realm of Boolean mathematics. Subtraction implies the existence of negative numbers: 5 – 3 is the same thing as 5 + (-3), and in Boolean algebra negative quantities are forbidden. There is no such thing as division in Boolean mathematics, either, since division is really nothing more than compounded subtraction, in the same way that multiplication is compounded addition.

Multiplication is valid in Boolean algebra, and thankfully it is the same as in real-number algebra: anything multiplied by 0 is 0, and anything multiplied by 1 remains unchanged:

$$0 \times 0 = 0$$

$$0 \times 1 = 0$$

$$1 \times 0 = 0$$

$$1 \times 1 = 1$$

This set of equations should also look familiar to you: it is the same pattern found in the truth table for an AND gate. In other words, Boolean multiplication corresponds to the logical function of an "AND" gate, as well as to series switch contacts:

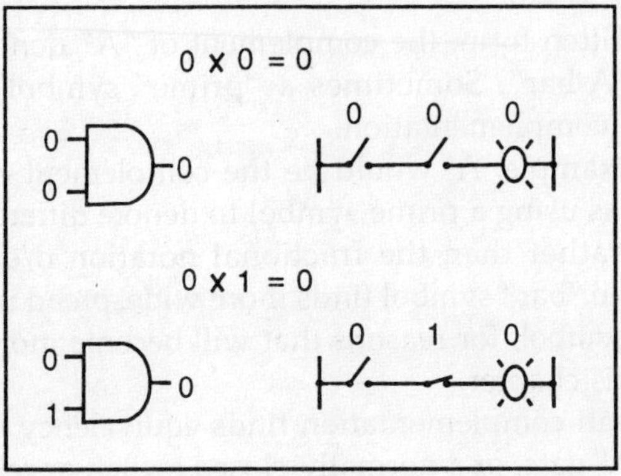

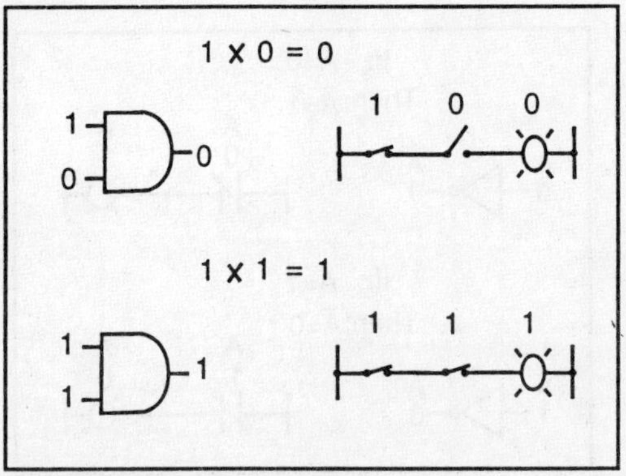

Like "normal" algebra, Boolean algebra uses alphabetical letters to denote variables. Unlike "normal" algebra, though, Boolean variables are always CAPITAL letters, never lower-case.

Because they are allowed to possess only one of two possible values, either 1 or 0, each and every variable has a *complement*: the opposite of its value. For example, if variable "A" has a value of 0, then the complement of A has a value of 1. Boolean notation uses a bar above the variable character to denote complementation, like this:

$$\text{If} : A = 0$$

$$\text{Then} : \bar{A} = 1$$

$$\text{If} : A = 0$$

$$\text{Then} : \bar{A} = 0$$

In written form, the complement of "A" denoted as "A-not" or "A-bar". Sometimes a "prime" symbol is used to represent complementation.

For example, A' would be the complement of A, much the same as using a prime symbol to denote differentiation in calculus rather than the fractional notation d/dt. Usually, though, the "bar" symbol finds more widespread use than the "prime" symbol, for reasons that will become more apparent later in this chapter.

Boolean complementation finds equivalency in the form of the NOT gate, or a normally-closed switch or relay contact:

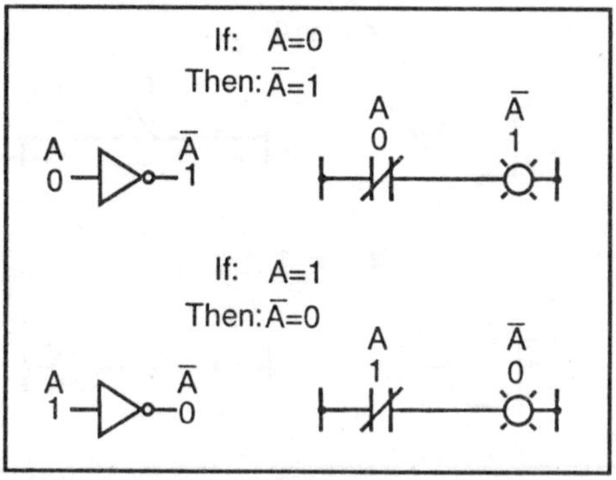

The basic definition of Boolean quantities has led to the simple rules of addition and multiplication, and has excluded both subtraction and division as valid arithmetic operations. We have a symbology for denoting Boolean variables, and their complements. In the next section we will proceed to develop Boolean identities.

REVIEW

- Boolean addition is equivalent to the *OR* logic function, as well as *parallel* switch contacts.

- Boolean multiplication is equivalent to the *AND* logic function, as well as *series* switch contacts.
- Boolean complementation is equivalent to the *NOT* logic function, as well as *normally-closed* relay contacts.

BOOLEAN ALGEBRAIC IDENTITIES

In mathematics, an *identity* is a statement true for all possible values of its variable or variables. The algebraic identity of x + 0 = x tells us that anything (x) added to zero equals the original "anything," no matter what value that "anything" (x) may be.

Like ordinary algebra, Boolean algebra has its own unique identities based on the bivalent states of Boolean variables.

The first Boolean identity is that the sum of anything and zero is the same as the original "anything." This identity is no different from its real-number algebraic equivalent:

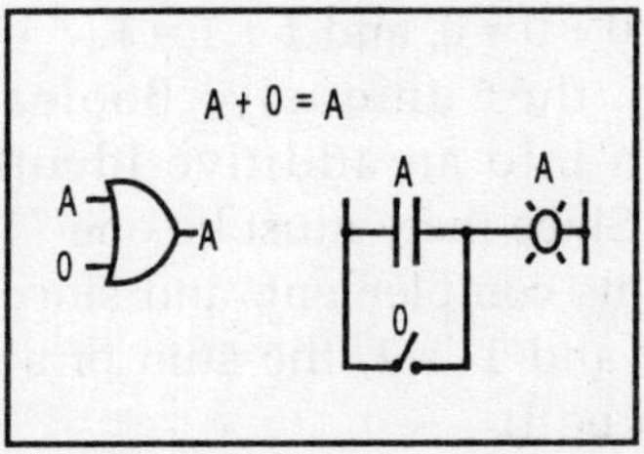

No matter what the value of A, the output will always be the same: when A=1, the output will also be 1; when A=0, the output will also be 0.

The next identity is most definitely *different* from any seen in normal algebra.

Here we discover that the sum of anything and one is one:

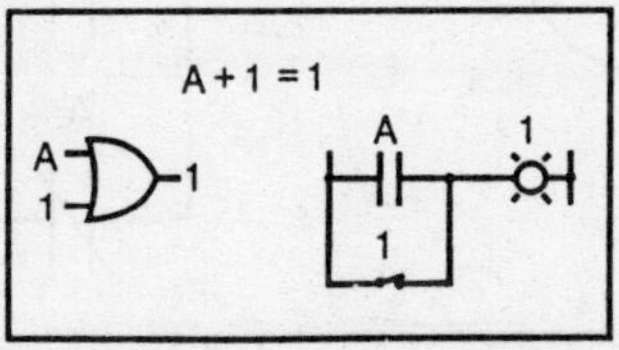

No matter what the value of A, the sum of A and 1 will always be 1. In a sense, the "1" signal *overrides* the effect of A on the logic circuit, leaving the output fixed at a logic level of 1. Next, we examine the effect of adding A and A together, which is the same as connecting both inputs of an OR gate to each other and activating them with the same signal:

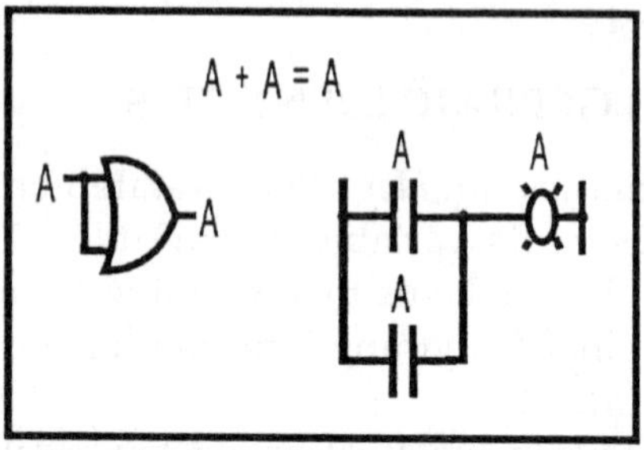

In real-number algebra, the sum of two identical variables is twice the original variable's value (x + x = 2x), but remember that there is no concept of "2" in the world of Boolean math, only 1 and 0, so we cannot say that A + A = 2A. Thus, when we add a Boolean quantity to itself, the sum is equal to the original quantity: 0 + 0 = 0, and 1 + 1 = 1.

Introducing the uniquely Boolean concept of complementation into an additive identity, we find an interesting effect. Since there must be one "1" value between any variable and its complement, and since the sum of any Boolean quantity and 1 is 1, the sum of a variable and its complement must be 1:

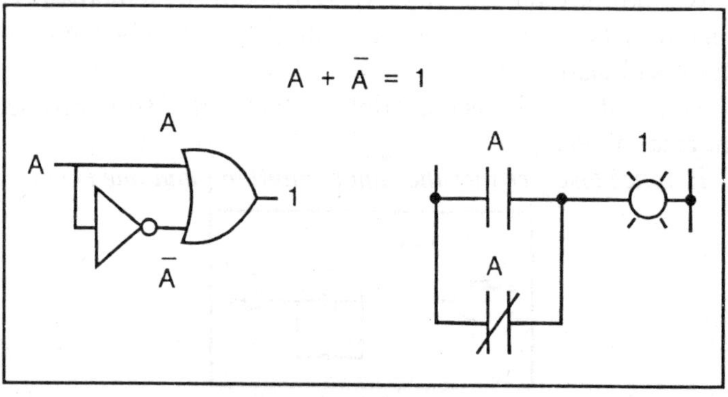

Just as there are four Boolean additive identities (A+0, A+1, A+A, and A+A'), so there are also four multiplicative identities:

A × 0, A × 1, A × A, and A × A′. Of these, the first two are no different from their equivalent expressions in regular algebra:

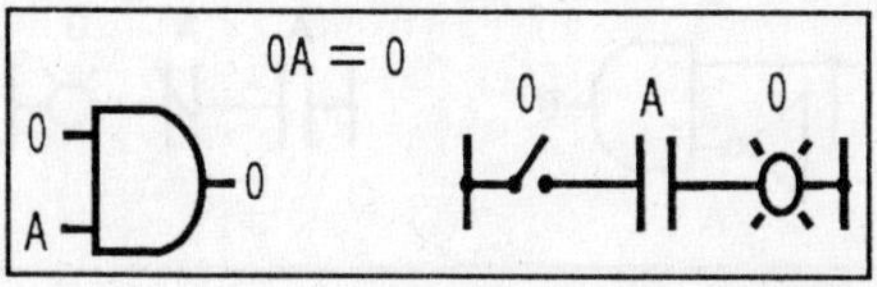

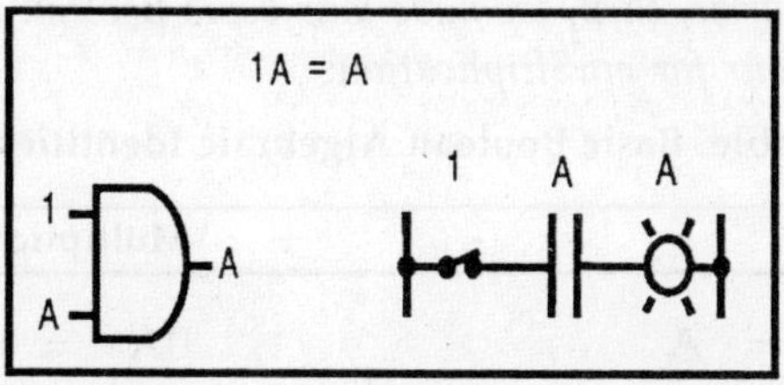

The third multiplicative identity expresses the result of a Boolean quantity multiplied by itself. In normal algebra, the product of a variable and itself is the *square* of that variable ($3 \times 3 = 3^2 = 9$).

However, the concept of "square" implies a quantity of 2, which has no meaning in Boolean algebra, so we cannot say that,

$$A \times A = A^2.$$

Instead, we find that the product of a Boolean quantity and itself is the original quantity, since,

$$0 \times 0 = 0 \text{ and } 1 \times 1 = 1:$$

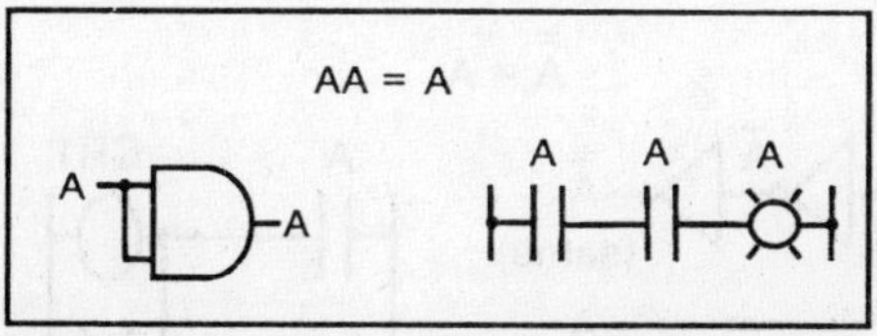

The fourth multiplicative identity has no equivalent in regular algebra because it uses the complement of a variable, a concept unique to Boolean mathematics.

Since there must be one "0" value between any variable and its complement, and since the product of any Boolean quantity and 0 is 0, the product of a variable and its complement must be 0:

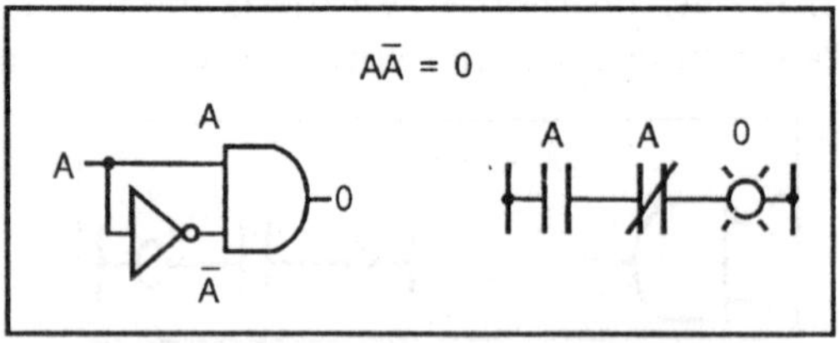

To summarize, then, we have four basic Boolean identities for addition and four for muSltiplication:

Table. Basic Boolean Algebraic Identities

Additive	Multiplicative
$A + 0 = A$	$0A = 0$
$A + 1 = 1$	$1A = A$
$A + A = A$	$AA = A$
$A + \overline{A} = 1$	$A\overline{A} = 0$

Another identity having to do with complementation is that of the *double complement*: a variable inverted twice. Complementing a variable twice (or any even number of times) results in the original Boolean value.

This is analogous to negating (multiplying by -1) in real-number algebra: an even number of negations cancel to leave the original value:

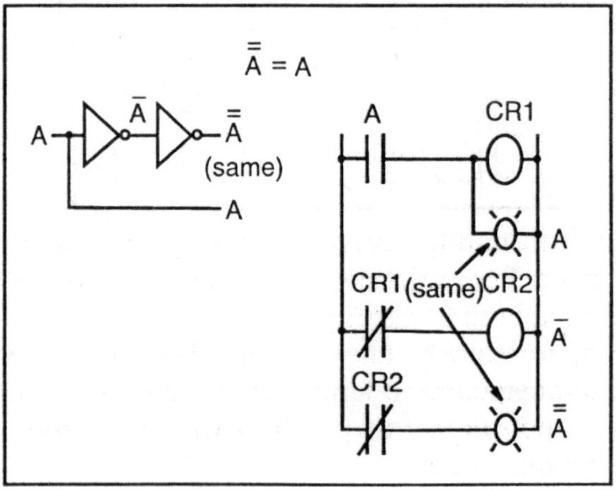

BOOLEAN ALGEBRAIC PROPERTIES

Another type of mathematical identity, called a "property" or a "law," describes how differing variables relate to each other in a system of numbers. One of these properties is known as the *commutative property*, and it applies equally to addition and multiplication. In essence, the commutative property tells us we can reverse the order of variables that are either added together or multiplied together without changing the truth of the expression:

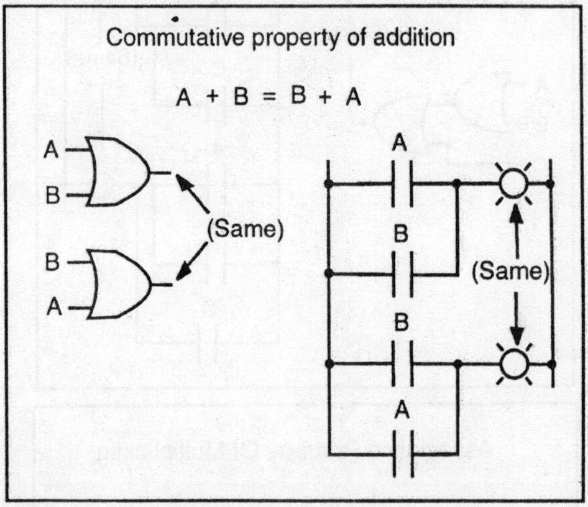

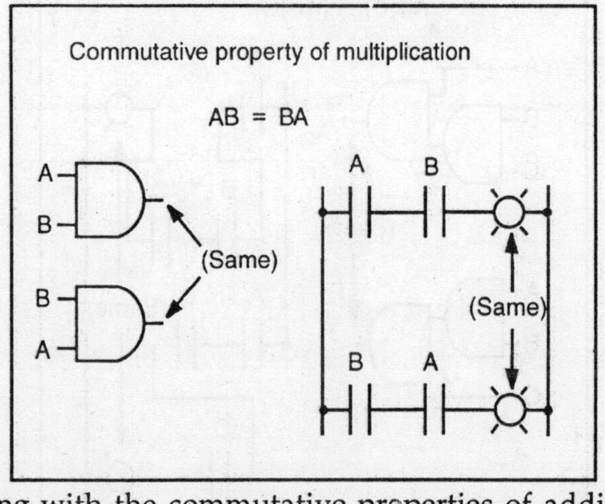

Along with the commutative properties of addition and multiplication, we have the *associative property*, again applying equally well to addition and multiplication. This property tells us we can associate groups of added or multiplied variables together with parentheses without altering the truth of the equations.

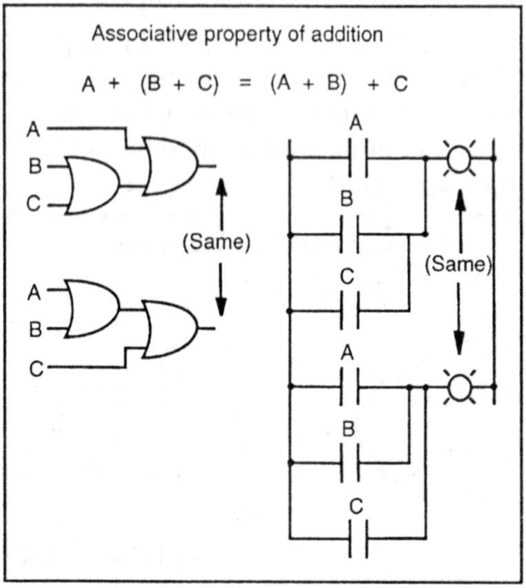

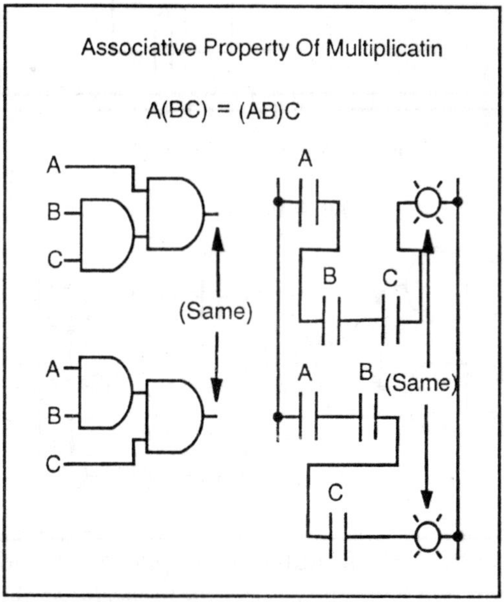

Lastly, we have the *distributive property*, illustrating how to expand a Boolean expression formed by the product of a sum, and in reverse shows us how terms may be factored out of Boolean sums-of-products:

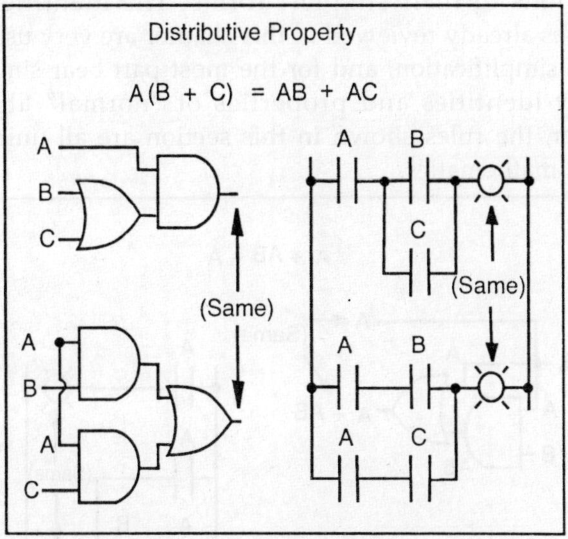

To summarize, here are the three basic properties: commutative, associative, and distributive.

Table.Basic Boolean Algebraic Properties

Additive	Multiplicative
A + B = B + A	AB = BA
A + (B + C) = (A + B) + C	A(BC) = (AB) C
A (B + C) = AB + AC	

BOOLEAN RULES FOR SIMPLIFICATION

Boolean algebra finds its most practical use in the simplification of logic circuits. If we translate a logic circuit's function into symbolic (Boolean) form, and apply certain algebraic rules to the resulting equation to reduce the number of terms and/or arithmetic operations, the simplified equation may be translated back into circuit form for a logic circuit performing the same function with fewer components. If equivalent function may be achieved with fewer components, the result will be increased reliability and decreased cost of manufacture. To this end, there are several rules of Boolean algebra presented in this section for use in reducing expressions to their simplest forms. The identities and properties already reviewed in this chapter are very useful in Boolean simplification, and for the most part bear similarity to many identities and properties of "normal" algebra. However, the rules shown in this section are all unique to Boolean mathematics.

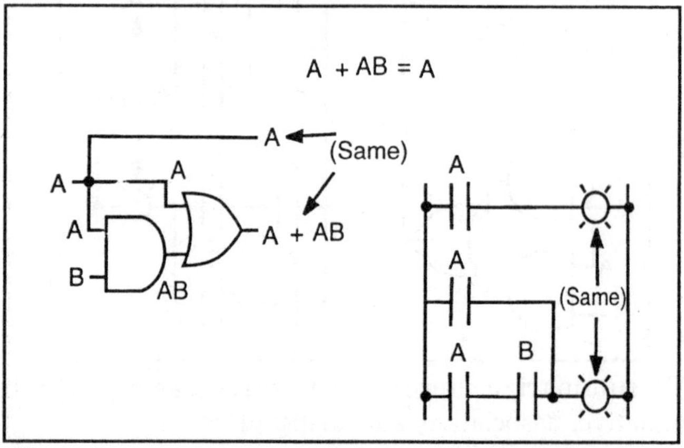

This rule may be proven symbolically by factoring an "A" out of the two terms, then applying the rules of A + 1 = 1 and 1A = A to achieve the final result:

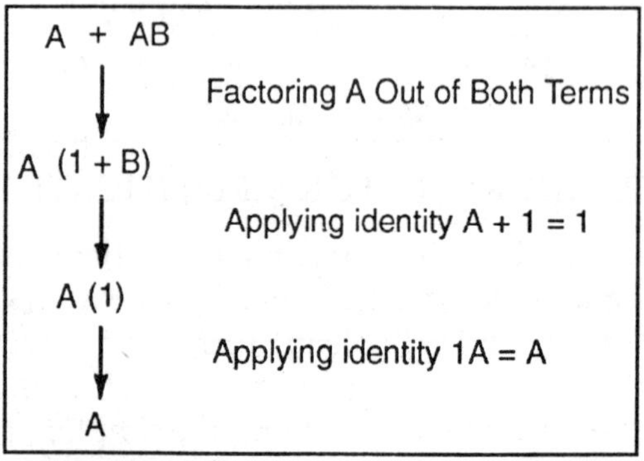

Please note how the rule A + 1 = 1 was used to reduce the (B + 1) term to 1. When a rule like "A + 1 = 1" is expressed using the letter "A", it doesn't mean it only applies to expressions containing "A". What the "A" stands for in a rule like A + 1 = 1 is *any* Boolean variable or collection of variables. This is perhaps the most difficult concept for new students to master in Boolean simplification: applying standardized identities, properties, and rules to expressions not in standard form. For instance, the Boolean expression ABC + 1 also reduces to 1 by means of the "A + 1 = 1" identity. In this case, we recognize that the "A" term in the identity's standard form can represent the entire "ABC" term in the original expression. The next rule looks similar to the first on shown in this section, but is actually quite different and requires a more clever proof:

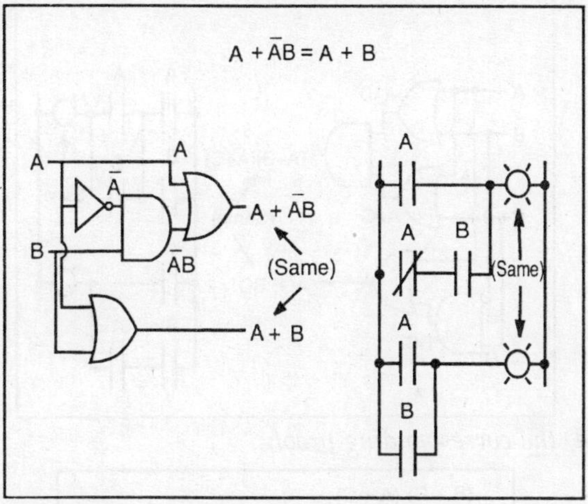

$A + \bar{A}B$

↓ Applying the Previous Rule to Expand A Term: A + AB = A

$A + AB + \bar{A}B$

↓ Factoring B Out of 2nd and 3rd Terms

$A + B(A + \bar{A})$

↓ Applying Identity A + A = 1

A + B (1)

↓ Applying Identity 1A = A

A + B

Note how the last rule (A + AB = A) is used to "unsimplify" the first "A" term in the expression, changing the "A" into an "A + AB".

While this may seem like a backward step, it certainly helped to reduce the expression to something simpler! Sometimes in mathematics we must take "backward" steps to achieve the most elegant solution. Knowing when to take such a step and when not to is part of the art-form of algebra, just as a victory in a game of chess almost always requires calculated sacrifices.

Another rule involves the simplification of a product-of-sums expression:

$$(A + B)(A + C) = A + BC$$

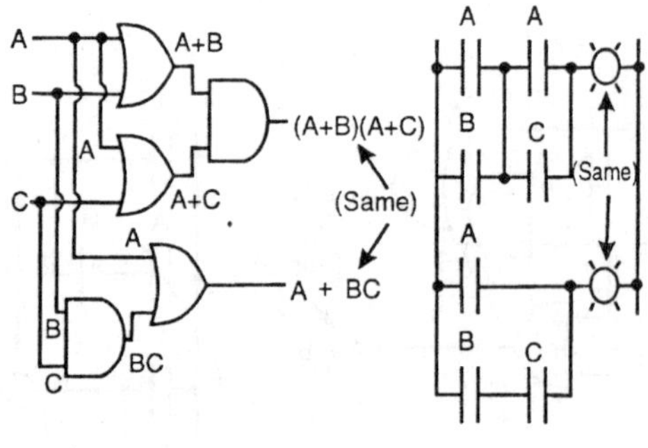

And, the corresponding proof:

(B + C) (A + C)

↓ Distributing Terms

AA + AC + AB + BC

↓ Applying Identity AA = A

A + AC + AB + BC

↓ Applying Rule A + AB = A to the A + AC term

A + AB + BC

↓ Applying Rule A + AB = A to the A + AB term

A + BC

To summarize, here are the three new rules of Boolean simplification expounded in this section:

$$A + AB = A$$

$$A + \bar{A}B = A + B$$

$$(A + B)(A + C) = A + BC$$

CIRCUIT SIMPLIFICATION EXAMPLES

Let's begin with a semiconductor gate circuit in need of simplification. The "A," "B," and "C" input signals are assumed to be provided from switches, sensors, or perhaps other gate circuits.

Where these signals originate is of no concern in the task of gate reduction.

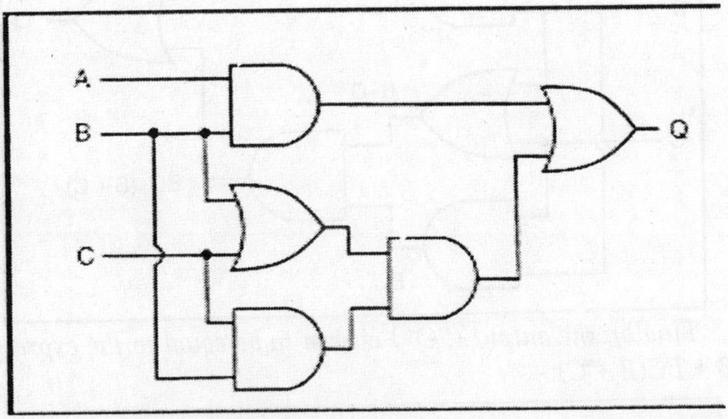

Our first step in simplification must be to write a Boolean expression for this circuit.

This task is easily performed step by step if we start by writing sub-expressions at the output of each gate, corresponding to the respective input signals for each gate.

Remember that OR gates are equivalent to Boolean addition, while AND gates are equivalent to Boolean multiplication.

For example, I'll write sub-expressions at the outputs of the first three gates:

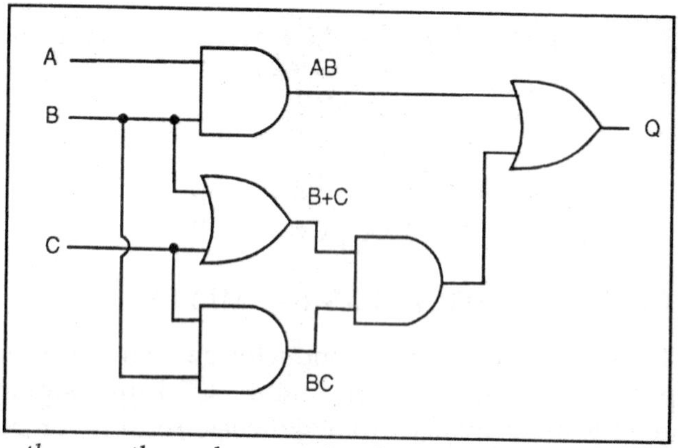

... then another sub-expression for the next gate:

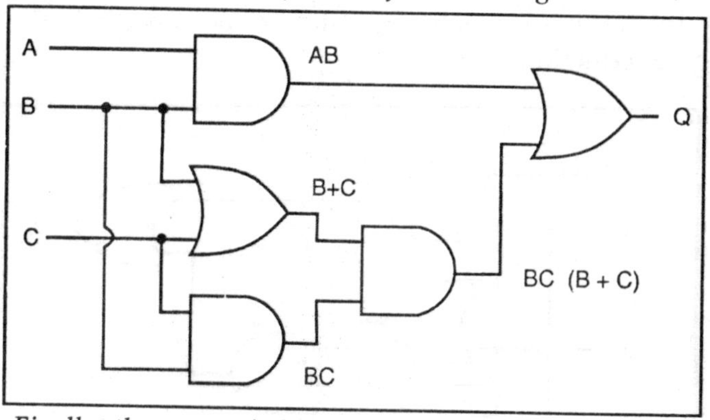

Finally, the output ("Q") is seen to be equal to the expression AB + BC(B + C):

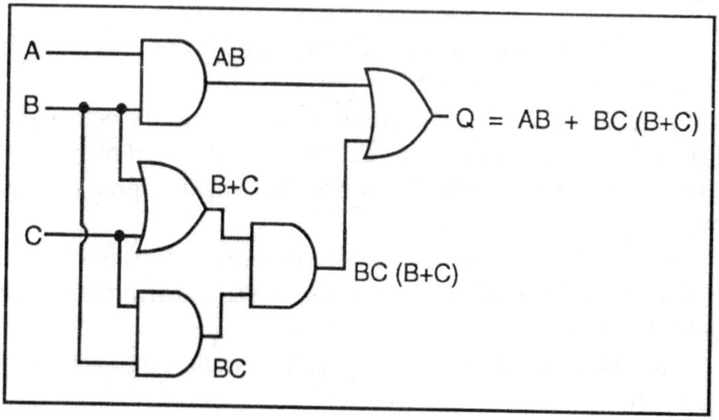

Now that we have a Boolean expression to work with, we need to apply the rules of Boolean algebra to reduce the expression to its simplest form (simplest defined as requiring the fewest gates to implement):

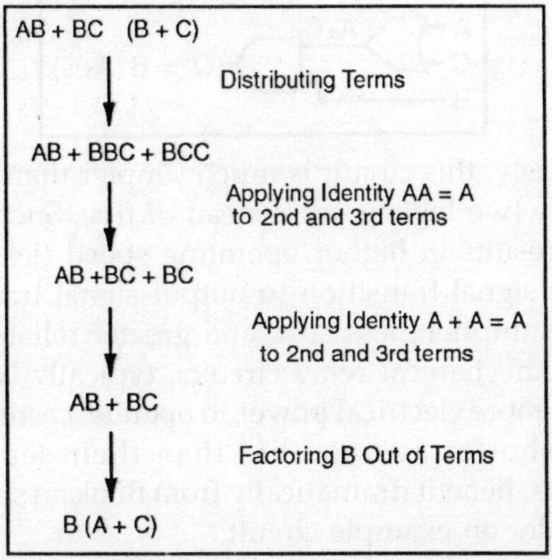

The final expression, B(A + C), is much simpler than the original, yet performs the same function. If you would like to verify this, you may generate a truth table for both expressions and determine Q's status (the circuits' output) for all eight logic-state combinations of A, B, and C, for both circuits. The two truth tables should be identical.

Now, we must generate a schematic diagram from this Boolean expression. To do this, evaluate the expression, following proper mathematical order of operations (multiplication before addition, operations inside parentheses before anything else), and draw gates for each step. Remember again that OR gates are equivalent to Boolean addition, while AND gates are equivalent to Boolean multiplication. In this case, we would begin with the sub-expression "A + C", which is an OR gate:

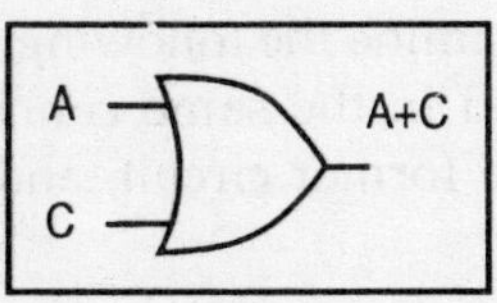

The next step in evaluating the expression "B(A + C)" is to multiply (AND gate) the signal B by the output of the previous gate (A + C):

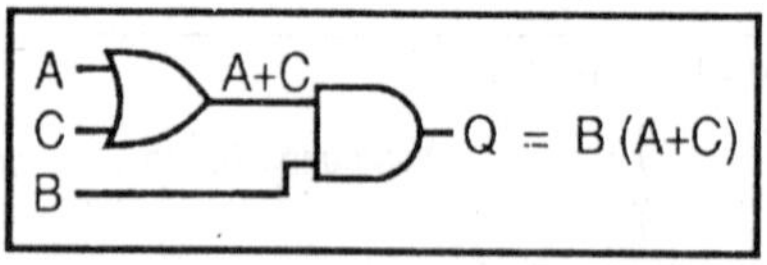

Obviously, this circuit is much simpler than the original, having only two logic gates instead of five. Such component reduction results in higher operating speed (less delay time from input signal transition to output signal transition), less power consumption, less cost, and greater reliability.

Electromechanical relay circuits, typically being slower, consuming more electrical power to operate, costing more, and having a shorter average life than their semiconductor counterparts, benefit dramatically from Boolean simplification. Let's consider an example circuit:

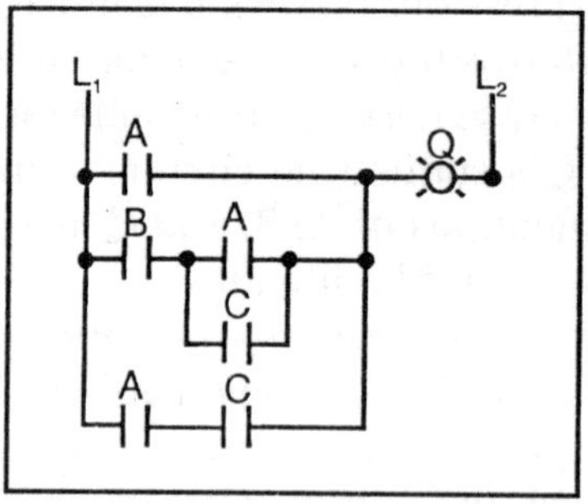

As before, our first step in reducing this circuit to its simplest form must be to develop a Boolean expression from the schematic.

The easiest way I've found to do this is to follow the same steps I'd normally follow to reduce a series-parallel resistor network to a single, total resistance.

For example, examine the following resistor network with its resistors arranged in the same connection pattern as the relay contacts in the former circuit, and corresponding total resistance formula:

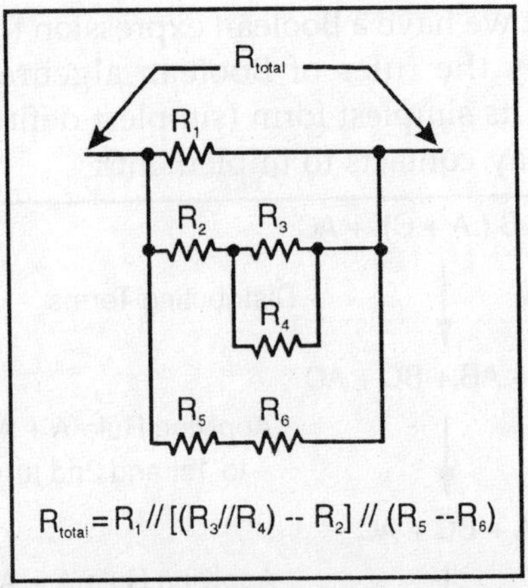

Remember that parallel contacts are equivalent to Boolean addition, while series contacts are equivalent to Boolean multiplication. Write a Boolean expression for this relay contact circuit, following the same order of precedence that you would follow in reducing a series-parallel resistor network to a total resistance. It may be helpful to write a Boolean sub-expression to the left of each ladder "rung," to help organize your expression-writing:

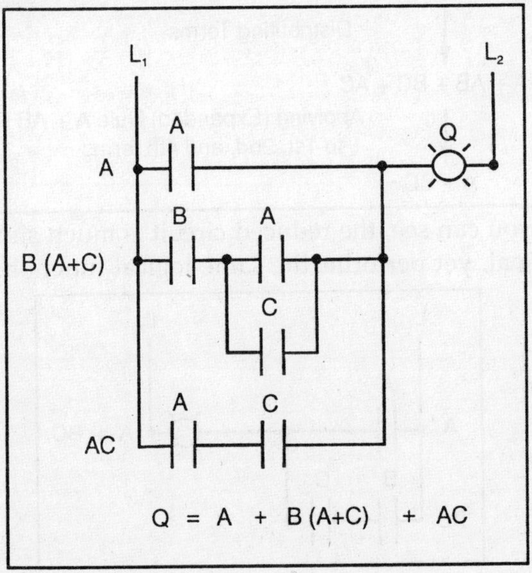

Now that we have a Boolean expression to work with, we need to apply the rules of Boolean algebra to reduce the expression to its simplest form (simplest defined as requiring the fewest relay contacts to implement):

$A + B(A + C) + AC$

↓ Distributing Terms

$A + AB + BC + AC$

↓ Applying Rule $A + AB = A$ to 1st and 2nd terms

$A + BC + AC$

↓ Applying Rule $A + AB = A$ to 1st and 3rd terms

$A + BC$

The more mathematically inclined should be able to see that the two steps employing the rule "A + AB = A" may be combined into a single step, the rule being expandable to: "A + AB + AC + AD +... = A"

$A + B(A + C) + AC$

↓ Distributing Terms

$A + AB + BC + AC$

↓ Applying (Expanded) Rule $A + AB = A$ to 1st, 2nd, and 4th terms

$A + BC$

As you can see, the reduced circuit is much simpler than the original, yet performs the same logical function:

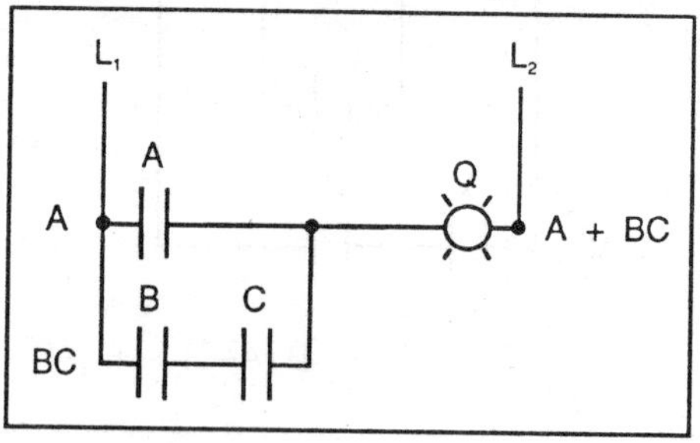

REVIEW

- To convert a gate circuit to a Boolean expression, label each gate output with a Boolean sub-expression corresponding to the gates' input signals, until a final expression is reached at the last gate.
- To convert a Boolean expression to a gate circuit, evaluate the expression using standard order of operations: multiplication before addition, and operations within parentheses before anything else.
- To convert a ladder logic circuit to a Boolean expression, label each rung with a Boolean sub-expression corresponding to the contacts' input signals, until a final expression is reached at the last coil or light. To determine proper order of evaluation, treat the contacts as though they were resistors, and as if you were determining total resistance of the series-parallel network formed by them. In other words, look for contacts that are either *directly* in series or *directly* in parallel with each other first, then "collapse" them into equivalent Boolean sub-expressions before proceeding to other contacts.
- To convert a Boolean expression to a ladder logic circuit, evaluate the expression using standard order of operations: multiplication before addition, and operations within parentheses before anything else.

EXCLUSIVE-OR FUNCTION

One element conspicuously missing from the set of Boolean operations is that of Exclusive-OR. Whereas the OR function is equivalent to Boolean addition, the AND function to Boolean multiplication, and the NOT function (inverter) to Boolean complementation, there is no direct Boolean equivalent for Exclusive-OR. This hasn't stopped people from developing a symbol to represent it, though:

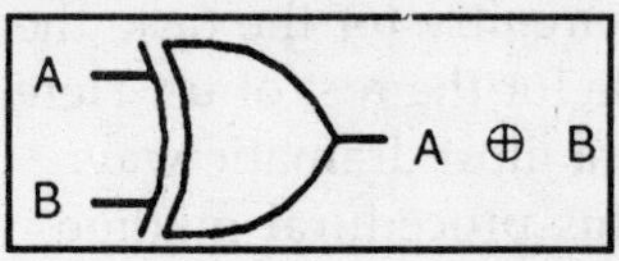

This symbol is seldom used in Boolean expressions because the identities, laws, and rules of simplification involving addition, multiplication, and complementation do not apply to it.

However, there is a way to represent the Exclusive-OR function in terms of OR and AND, as has been shown in previous chapters:

AB′ + A′B

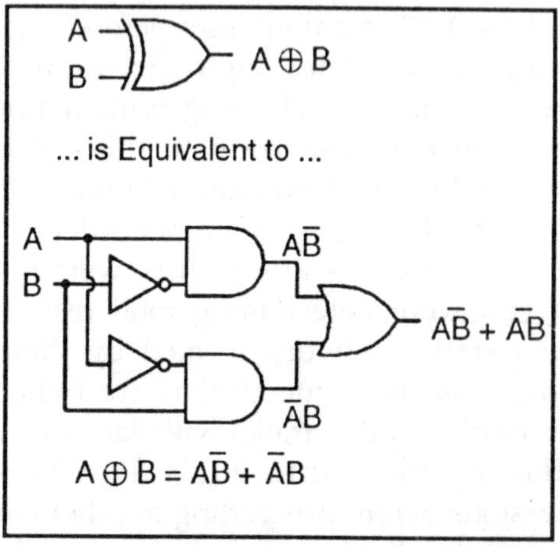

As a Boolean equivalency, this rule may be helpful in simplifying some Boolean expressions. Any expression following the AB′ + A′B form (two AND gates and an OR gate) may be replaced by a single Exclusive-OR gate.

CONVERTING TRUTH TABLES INTO BOOLEAN EXPRESSIONS

In designing digital circuits, the designer often begins with a truth table describing what the circuit should do. The design task is largely to determine what type of circuit will perform the function described in the truth table.

While some people seem to have a natural ability to look at a truth table and immediately envision the necessary logic gate or relay logic circuitry for the task, there are procedural techniques available for the rest of us. Here, Boolean algebra proves its utility in a most dramatic way.

To illustrate this procedural method, we should begin

with a realistic design problem. Suppose we were given the task of designing a flame detection circuit for a toxic waste incinerator.

The intense heat of the fire is intended to neutralize the toxicity of the waste introduced into the incinerator. Such combustion-based techniques are commonly used to neutralize medical waste, which may be infected with deadly viruses or bacteria:

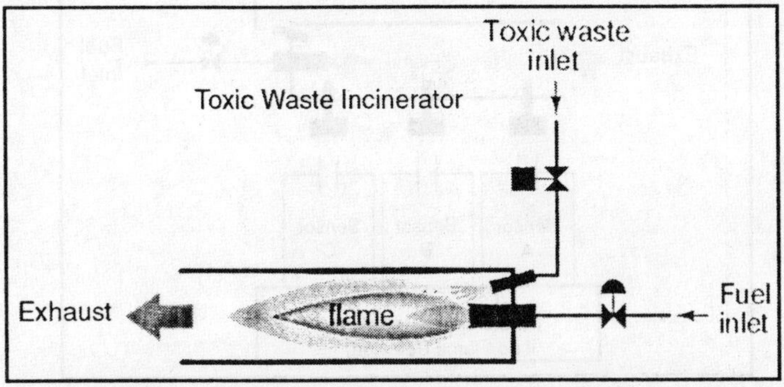

So long as a flame is maintained in the incinerator, it is safe to inject waste into it to be neutralized. If the flame were to be extinguished, however, it would be unsafe to continue to inject waste into the combustion chamber, as it would exit the exhaust un-neutralized, and pose a health threat to anyone in close proximity to the exhaust.

What we need in this system is a sure way of detecting the presence of a flame, and permitting waste to be injected only if a flame is "proven" by the flame detection system.

Several different flame-detection technologies exist: optical (detection of light), thermal (detection of high temperature), and electrical conduction (detection of ionized particles in the flame path), each one with its unique advantages and disadvantages.

Suppose that due to the high degree of hazard involved with potentially passing un-neutralized waste out the exhaust of this incinerator, it is decided that the flame detection system be made redundant (multiple sensors), so that failure of a single sensor does not lead to an emission of toxins out the exhaust. Each sensor comes equipped with a normally-open contact (open if no flame, closed if flame detected) which we will use to activate the inputs of a logic system:

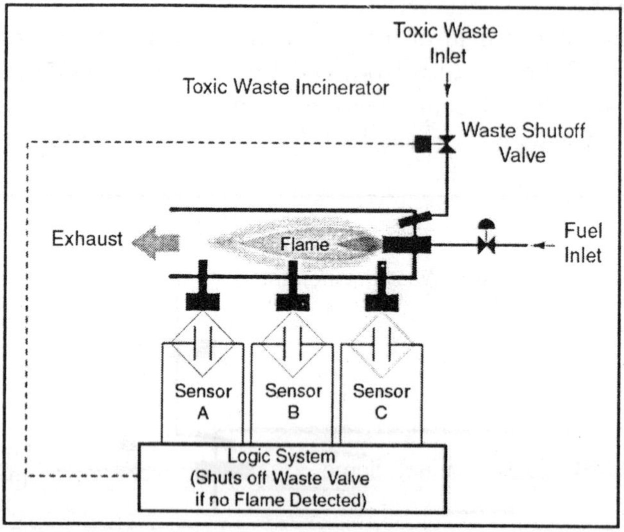

Our task, now, is to design the circuitry of the logic system to open the waste valve if and only if there is good flame proven by the sensors. First, though, we must decide what the logical behaviour of this control system should be. Do we want the valve to be opened if only one out of the three sensors detects flame? Probably not, because this would defeat the purpose of having multiple sensors.

If any one of the sensors were to fail in such a way as to falsely indicate the presence of flame when there was none, a logic system based on the principle of "any one out of three sensors showing flame" would give the same output that a single-sensor system would with the same failure. A far better solution would be to design the system so that the valve is commanded to open if any only if *all three sensors* detect a good flame. This way, any single, failed sensor falsely showing flame could not keep the valve in the open position; rather, it would require all three sensors to be failed in the same manner — a highly improbable scenario — for this dangerous condition to occur.

Thus, our truth table would look like this:

Sensor Inputs

A	B	C	Output
0	0	0	0
0	0	1	0
0	1	0	0
0	1	1	0
1	0	0	0
1	0	1	0
1	1	0	0
1	1	1	1

Output = 0 (Close Valve)

Output = 1 (Open Valve)

It does not require much insight to realise that this functionality could be generated with a three-input AND gate: the output of the circuit will be "high" if and only if input A *AND* input B *AND* input C are all "high:"

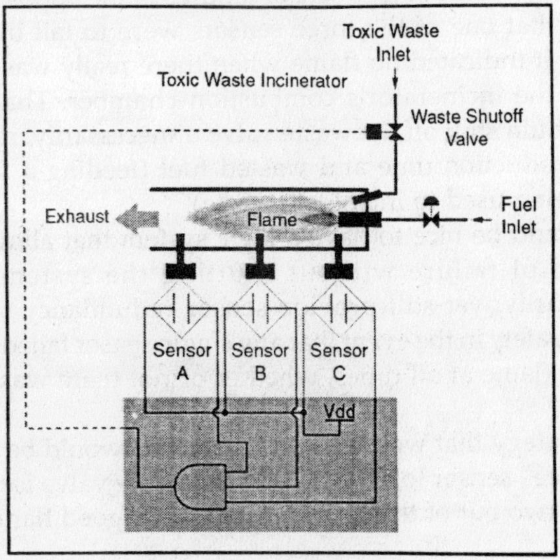

If using relay circuitry, we could create this AND function by wiring three relay contacts in series, or simply by wiring the three sensor contacts in series, so that the only way electrical power could be sent to open the waste valve is if all three sensors indicate flame:

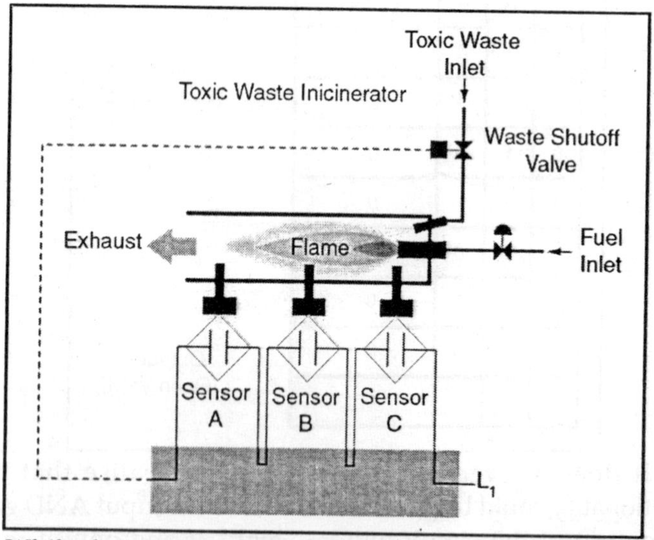

While this design strategy maximizes safety, it makes the system very susceptible to sensor failures of the opposite kind. Suppose that one of the three sensors were to fail in such a way that it indicated no flame when there really was a good flame in the incinerator's combustion chamber. That single failure would shut off the waste valve unnecessarily, resulting in lost production time and wasted fuel (feeding a fire that wasn't being used to incinerate waste).

It would be nice to have a logic system that allowed for this kind of failure without shutting the system down unnecessarily, yet still provide sensor redundancy so as to maintain safety in the event that any single sensor failed "high" (showing flame at all times, whether or not there was one to detect).

A strategy that would meet both needs would be a "two out of three" sensor logic, whereby the waste valve is opened if at least two out of the three sensors show good flame.

The truth table for such a system would look like this:

Sensor Inputs

A	B	C	Output
0	0	0	0
0	0	1	0
0	1	0	0
0	1	1	1
1	0	0	0
1	0	1	1
1	1	0	1
1	1	1	1

Output = 0 (Close Valve)

Output = 1 (Open Valve)

Here, it is not necessarily obvious what kind of logic circuit would satisfy the truth table. However, a simple method for designing such a circuit is found in a standard form of Boolean expression called the *Sum-Of-Products*, or *SOP*, form. As you might suspect, a Sum-Of-Products Boolean expression is literally a set of Boolean terms added (*summed*) together, each term being a multiplicative (*product*) combination of Boolean variables.

An example of an SOP expression would be something like this: ABC + BC + DF, the sum of products "ABC," "BC," and "DF."

Sum-Of-Products expressions are easy to generate from truth tables. All we have to do is examine the truth table for any rows where the output is "high" (1), and write a Boolean product term that would equal a value of 1 given those input conditions. For instance, in the fourth row down in the truth table for our two-out-of-three logic system, where A=0, B=1, and C=1, the product term would be A'BC, since that term would have a value of 1 if and only if A=0, B=1, and C=1:

Sensor Inputs

A	B	C	Output	
0	0	0	0	
0	0	1	0	
0	1	0	0	
0	1	1	1	$\bar{A}BC = 1$
1	0	0	0	
1	0	1	1	
1	1	0	1	
1	1	1	1	

Three other rows of the truth table have an output value of 1, so those rows also need Boolean product expressions to represent them:

Sensor Inputs

A	B	C	Output	
0	0	0	0	
0	0	1	0	
0	1	0	0	
0	1	1	1	$\bar{A}BC = 1$
1	0	0	0	
1	0	1	1	$A\bar{B}C = 1$
1	1	0	1	$AB\bar{C} = 1$
1	1	1	1	$ABC = 1$

Finally, we join these four Boolean product expressions together by addition, to create a single Boolean expression describing the truth table as a whole:

Sensor Inputs

A	B	C	Output	
0	0	0	0	
0	0	1	0	
0	1	0	0	
0	1	1	1	$\overline{A}BC = 1$
1	0	0	0	
1	0	1	1	$A\overline{B}C = 1$
1	1	0	1	$AB\overline{C} = 1$
1	1	1	1	$ABC = 1$

$$\text{Output} = \overline{A}BC + A\overline{B}C + AB\overline{C} + ABC$$

Now that we have a Boolean Sum-Of-Products expression for the truth table's function, we can easily design a logic gate or relay logic circuit based on that expression:

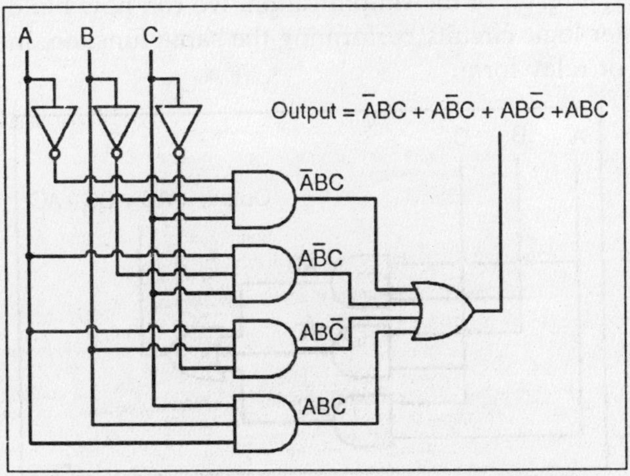

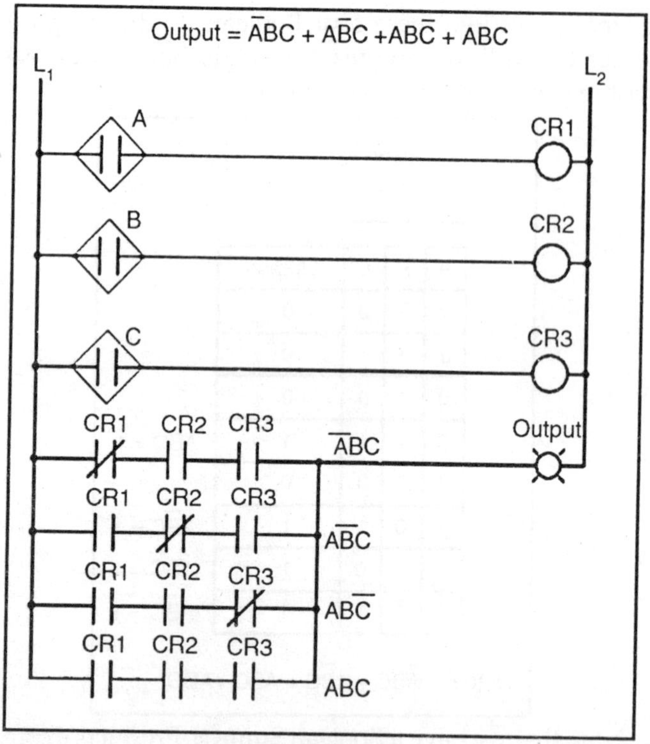

Unfortunately, both of these circuits are quite complex, and could benefit from simplification.

The result of the simplification, we can now build much simpler logic circuits performing the same function, in either gate or relay form:

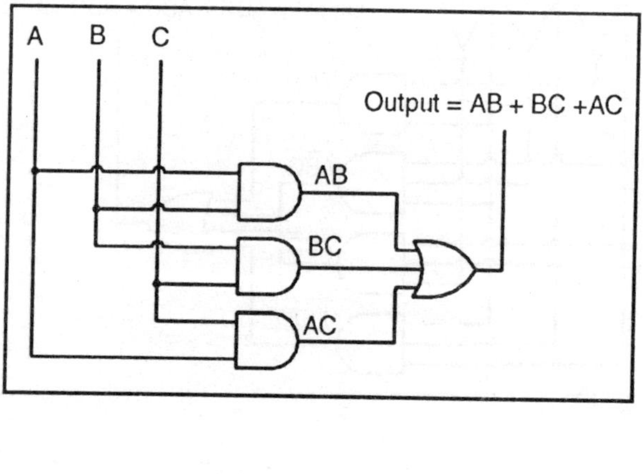

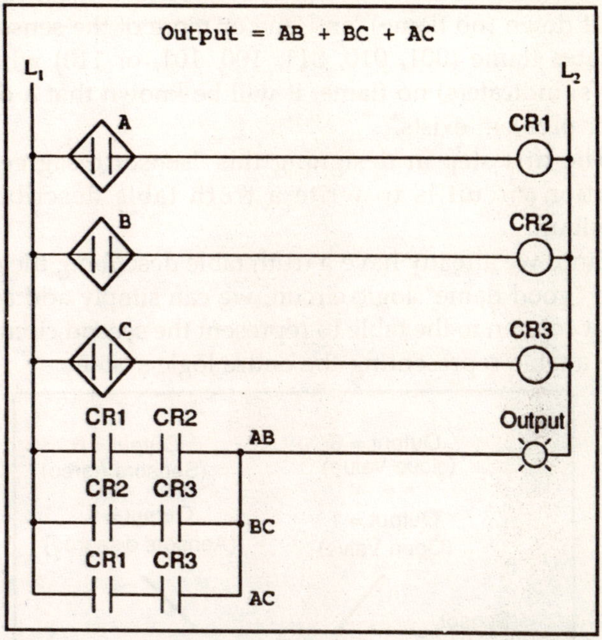

Either one of these circuits will adequately perform the task of operating the incinerator waste valve based on a flame verification from two out of the three flame sensors. At minimum, this is what we need to have a safe incinerator system. We can, however, extend the functionality of the system by adding to it logic circuitry designed to detect if any one of the sensors does not agree with the other two.

If all three sensors are operating properly, they should detect flame with equal accuracy. Thus, they should either all register "low" (000: no flame) or all register "high" (111: good flame). Any other output combination (001, 010, 011, 100, 101, or 110) constitutes a disagreement between sensors, and may therefore serve as an indicator of a potential sensor failure. If we added circuitry to detect any one of the six "sensor disagreement" conditions, we could use the output of that circuitry to activate an alarm.

Whoever is monitoring the incinerator would then exercise judgment in either continuing to operate with a possible failed sensor (inputs: 011, 101, or 110), or shut the incinerator down to be absolutely safe. Also, if the incinerator is shut down (no flame), and one or more of the sensors still indicates flame (001, 010, 011, 100, 101, or 110) while the other(s) indicate(s) no flame, it will be known that a definite sensor problem exists.

The first step in designing this "sensor disagreement" detection circuit is to write a truth table describing its behaviour.

Since we already have a truth table describing the output of the "good flame" logic circuit, we can simply add another output column to the table to represent the second circuit, and make a table representing the entire logic system:

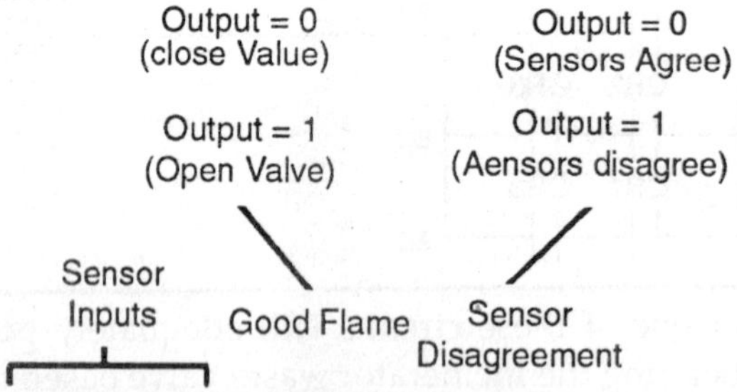

A	B	C	Output	Output
0	0	0	0	0
0	0	1	0	1
0	1	0	0	1
0	1	1	1	1
1	0	0	0	1
1	0	1	1	1
1	1	0	1	1
1	1	1	1	0

While it is possible to generate a Sum-Of-Products expression for this new truth table column, it would require six terms, of three variables each! Such a Boolean expression would require many steps to simplify, with a large potential for making algebraic errors:

Output = 0 (Close Valve)
Output = 1 (Open Valve)

Output = 0 (Sensors Agree)
Output = 1 (Sensors Disagree)

Sensor Inputs			Good Flame	Sensor Disagreement	
A	B	C	Output	Output	
0	0	0	0	0	
0	0	1	0	1	$\bar{A}\bar{B}C$
0	1	0	0	1	$\bar{A}B\bar{C}$
0	1	1	1	1	$\bar{A}BC$
1	0	0	0	1	$A\bar{B}\bar{C}$
1	0	1	1	1	$A\bar{B}C$
1	1	0	1	1	$AB\bar{C}$
1	1	1	1	0	

$$\text{Output} = \bar{A}\bar{B}C + \bar{A}B\bar{C} + \bar{A}BC + A\bar{B}C + A\bar{B}C + AB\bar{C}$$

An alternative to generating a Sum-Of-Products expression to account for all the "high" (1) output conditions in the truth table is to generate a *Product-Of-Sums*, or *POS*, expression, to account for all the "low" (0) output conditions instead.

Being that there are much fewer instances of a "low" output in the last truth table column, the resulting Product-Of-Sums expression should contain fewer terms. As its name suggests, a Product-Of-Sums expression is a set of added terms (*sums*), which are multiplied (*product*) together. An example of a POS expression would be (A + B)(C + D), the product of the sums "A + B" and "C + D".

To begin, we identify which rows in the last truth table column have "low" (0) outputs, and write a Boolean sum term that would equal 0 for that row's input conditions. For instance, in the first row of the truth table, where A=0, B=0, and C=0, the sum term would be (A + B + C), since that term would have a value of 0 if and only if A=0, B=0, and C=0:

Output = 0 (Close Valve) / Output = 1 (Open Valve) — Good Flame

Output = 0 (Sensors Agree) / Output = 1 (Sensors Disagree) — Sensor Disagreement

Sensor Inputs A	B	C	Good Flame Output	Sensor Disagreement Output	
0	0	0	0	0	(A + B + C)
0	0	1	0	1	
0	1	0	0	1	
0	1	1	1	1	
1	0	0	0	1	
1	0	1	1	1	
1	1	0	1	1	
1	1	1	1	0	

Only one other row in the last truth table column has a "low" (0) output, so all we need is one more sum term to complete our Product-Of-Sums expression. This last sum term represents a 0 output for an input condition of A=1, B=1 and C=1. Therefore, the term must be written as (A′ + B′+ C′), because only the sum of the *complemented* input variables would equal 0 for that condition only:

Output = 0 (Close Valve) / Output = 1 (Open Valve) — Good Flame

Output = 0 (Sensors Agree) / Output = 1 (Sensors Disagree) — Sensor Disagreement

Sensor Inputs A	B	C	Good Flame Output	Sensor Disagreement Output	
0	0	0	0	0	(A + B + C)
0	0	1	0	1	
0	1	0	0	1	
0	1	1	1	1	
1	0	0	0	1	
1	0	1	1	1	
1	1	0	1	1	
1	1	1	1	0	$(\bar{A} + \bar{B} + \bar{C})$

The completed Product-Of-Sums expression, of course, is the multiplicative combination of these two sum terms:

Output = 0 (Close Valve) Output = 1 (Open Valve) — Good Flame

Output = 0 (Sensors Agree) Output = 1 (Sensors Disagree) — Sensor Disagreement

Sensor Inputs

A	B	C	Good Flame Output	Sensor Disagreement Output	
0	0	0	0	0	$(A + B + C)$
0	0	1	0	1	
0	1	0	0	1	
0	1	1	1	1	
1	0	0	0	1	
1	0	1	1	1	
1	1	0	1	1	
1	1	1	1	0	$(\bar{A} + \bar{B} + \bar{C})$

$\text{Output} = (A + B + C)(\bar{A} + \bar{B} + C)$

Whereas a Sum-Of-Products expression could be implemented in the form of a set of AND gates with their outputs connecting to a single OR gate, a Product-Of-Sums expression can be implemented as a set of OR gates feeding into a single AND gate:

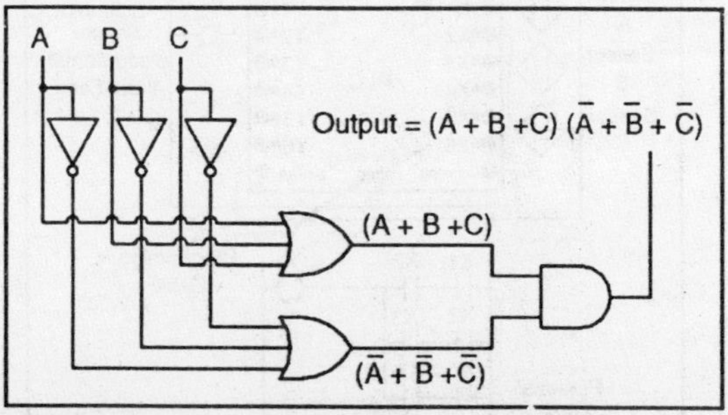

Correspondingly, whereas a Sum-Of-Products expression could be implemented as a parallel collection of series-connected relay contacts, a Product-Of-Sums expression can be implemented as a series collection of parallel-connected relay contacts:

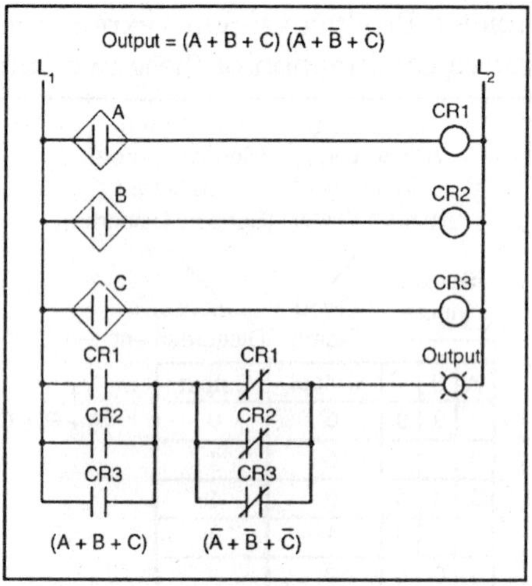

The previous two circuits represent different versions of the "sensor disagreement" logic circuit only, not the "good flame" detection circuit(s). The entire logic system would be the combination of both "good flame" and "sensor disagreement" circuits, shown on the same diagram.

Implemented in a Programmable Logic Controller (PLC), the entire logic system might resemble something like this:

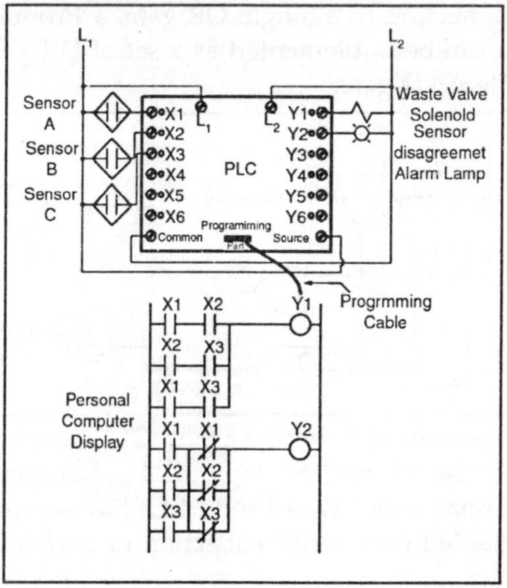

As you can see, both the Sum-Of-Products and Products-Of-Sums standard Boolean forms are powerful tools when applied to truth tables. They allow us to derive a Boolean expression — and ultimately, an actual logic circuit — from nothing but a truth table, which is a written specification for what we want a logic circuit to do. To be able to go from a written specification to an actual circuit using simple, deterministic procedures means that it is possible to automate the design process for a digital circuit.

In other words, a computer could be programmed to design a custom logic circuit from a truth table specification! The steps to take from a truth table to the final circuit are so unambiguous and direct that it requires little, if any, creativity or other original thought to execute them.

Chapter 3

Logic Gate Family

DIGITAL SIGNALS AND GATES

While the binary numeration system is an interesting mathematical abstraction, we haven't yet seen its practical application to electronics. This chapter is devoted to just that: practically applying the concept of binary bits to circuits. What makes binary numeration so important to the application of digital electronics is the ease in which bits may be represented in physical terms.

Because a binary bit can only have one of two different values, either 0 or 1, any physical medium capable of switching between two saturated states may be used to represent a bit. Consequently, any physical system capable of representing binary bits is able to represent numerical quantities, and potentially has the ability to manipulate those numbers. This is the basic concept underlying digital computing.

Electronic circuits are physical systems that lend themselves well to the representation of binary numbers. Transistors, when operated at their bias limits, may be in one of two different states: either cutoff (no controlled current) or saturation (maximum controlled current). If a transistor circuit is designed to maximize the probability of falling into either one of these states (and not operating in the linear, or *active,* mode), it can serve as a physical representation of a binary bit. A voltage signal measured at the output of such a circuit may also serve as a representation of a single bit, a low voltage representing a binary "0" and a (relatively) high voltage representing a binary "1."

Note the following transistor circuit:

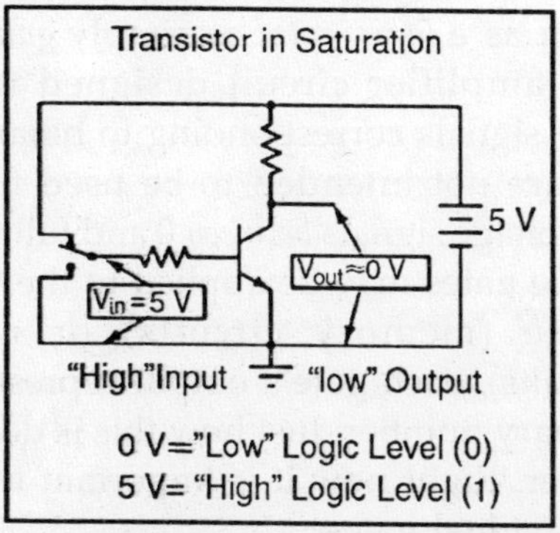

In this circuit, the transistor is in a state of saturation by virtue of the applied input voltage (5 volts) through the two-position switch. Because it's saturated, the transistor drops very little voltage between collector and emitter, resulting in an output voltage of (practically) 0 volts. If we were using this circuit to represent binary bits, we would say that the input signal is a binary "1" and that the output signal is a binary "0." Any voltage close to full supply voltage (measured in reference to ground, of course) is considered a "1" and a lack of voltage is considered a "0." Alternative terms for these voltage levels are*high* (same as a binary "1") and *low* (same as a binary "0"). A general term for the representation of a binary bit by a circuit voltage is *logic level.*

Moving the switch to the other position, we apply a binary "0" to the input and receive a binary "1" at the output:

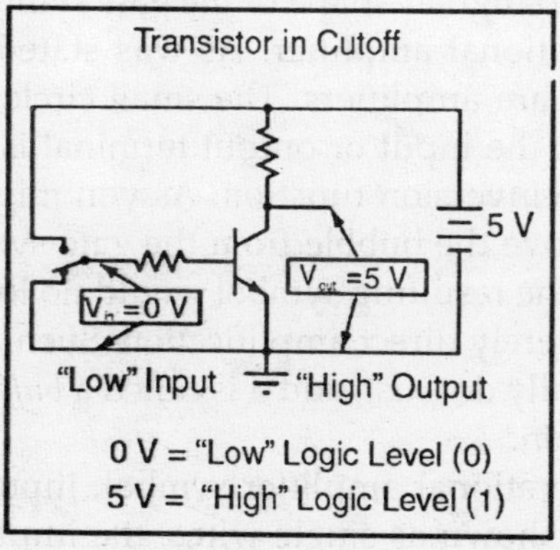

What we've created here with a single transistor is a circuit generally known as a *logic gate*, or simply *gate*. A gate is a special type of amplifier circuit designed to accept and generate voltage signals corresponding to binary 1's and 0's. As such, gates are not intended to be used for amplifying analog signals (voltage signals *between* 0 and full voltage). Used together, multiple gates may be applied to the task of binary number storage (memory circuits) or manipulation (computing circuits), each gate's output representing one bit of a multi-bit binary number. Just how this is done is a subject for a later chapter. Right now it is important to focus on the operation of individual gates.

The gate shown here with the single transistor is known as an *inverter*, or NOT gate, because it outputs the exact opposite digital signal as what is input. For convenience, gate circuits are generally represented by their own symbols rather than by their constituent transistors and resistors.

The following is the symbol for an inverter:

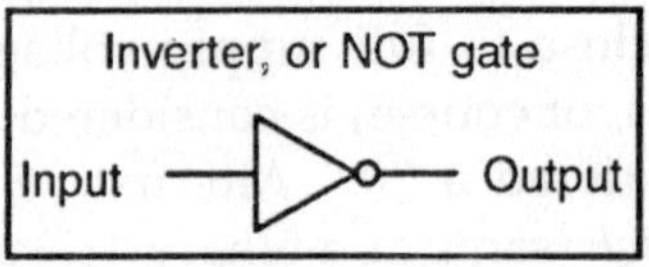

An alternative symbol for an inverter is shown here:

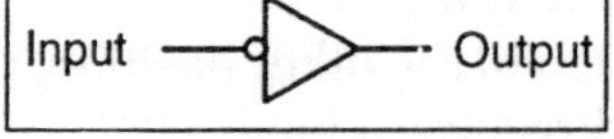

Notice the triangular shape of the gate symbol, much like that of an operational amplifier. As was stated before, gate circuits actually are amplifiers. The small circle, or "bubble" shown on either the input or output terminal is standard for representing the inversion function. As you might suspect, if we were to remove the bubble from the gate symbol, leaving only a triangle, the resulting symbol would no longer indicate inversion, but merely direct amplification. Such a symbol and such a gate actually do exist, and it is called a *buffer*, the subject of the next section.

Like an operational amplifier symbol, input and output connections are shown as single wires, the implied reference point for each voltage signal being "ground." In digital gate circuits, ground is almost always the negative connection of a single voltage source (power supply). Dual, or "split," power supplies are seldom used in gate circuitry. Because gate circuits are amplifiers, they require a source of power to operate. Like operational amplifiers, the power supply connections for digital gates are often omitted from the symbol for simplicity's sake. If we were to show *all* the necessary connections needed for operating this gate, the schematic would look something like this:

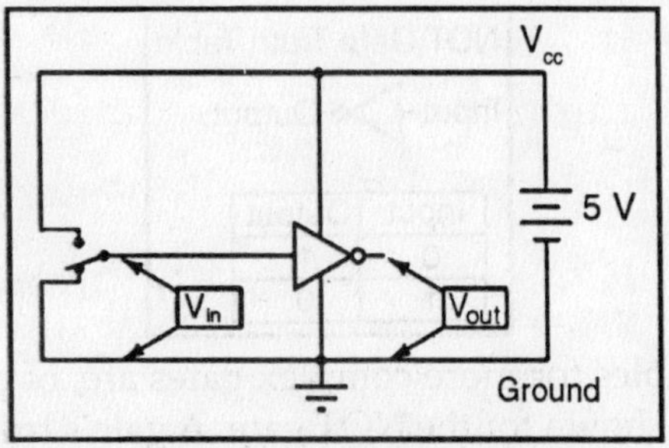

Power supply conductors are rarely shown in gate circuit schematics, even if the power supply connections at each gate are. *Minimizing lines in our schematic, we get this*:

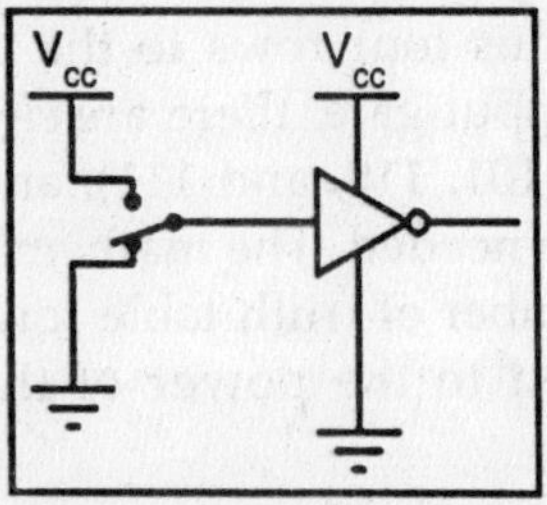

"V_{cc}" stands for the constant voltage supplied to the collector of a bipolar junction transistor circuit, in reference to ground. Those points in a gate circuit marked by the label "V_{cc}" are all connected to the same point, and that point is the positive terminal of a DC voltage source, usually 5 volts.

As we will see in other sections of this chapter, there are quite a few different types of logic gates, most of which have multiple input terminals for accepting more than one signal. The output of any gate is dependent on the state of its input(s) and its logical function.

One common way to express the particular function of a gate circuit is called a *truth table*. Truth tables show all combinations of input conditions in terms of logic level states (either "high" or "low," "1" or "0," for each input terminal of the gate), along with the corresponding output logic level, either "high" or "low." For the inverter, or NOT, circuit just illustrated, the truth table is very simple indeed:

NOT Gate Truth Table

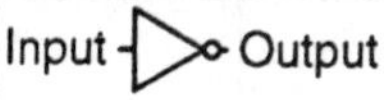

Input	Output
0	1
1	0

Truth tables for more complex gates are, of course, larger than the one shown for the NOT gate. A gate's truth table must have as many rows as there are possibilities for unique input combinations. For a single-input gate like the NOT gate, there are only two possibilities, 0 and 1.

For a two input gate, there are *four* possibilities (00, 01, 10, and 11), and thus four rows to the corresponding truth table. For a three-input gate, there are *eight* possibilities (000, 001, 010, 011, 100, 101, 110, and 111), and thus a truth table with eight rows are needed. The mathematically inclined will realise that the number of truth table rows needed for a gate is equal to 2 raised to the power of the number of input terminals.

THE NOT GATE

The single-transistor inverter circuit illustrated earlier is actually too crude to be of practical use as a gate. Real inverter circuits contain more than one transistor to maximize voltage gain (so as to ensure that the final output transistor is either in full cutoff or full saturation), and other components designed to reduce the chance of accidental damage.

Shown here is a schematic diagram for a real inverter circuit, complete with all necessary components for efficient and reliable operation:

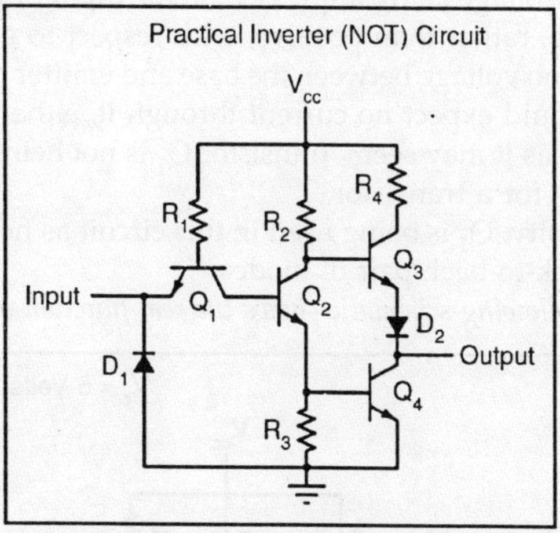

This circuit is composed exclusively of resistors and bipolar transistors. Bear in mind that other circuit designs are capable of performing the NOT gate function, including designs substituting field-effect transistors for bipolar (discussed later in this chapter).

Let's analyse this circuit for the condition where the input is "high," or in a binary "1" state. We can simulate this by showing the input terminal connected to V_{cc} through a switch:

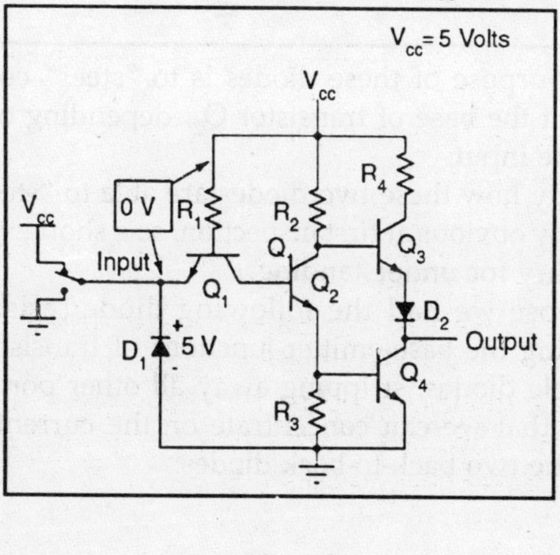

In this case, diode D_1 will be reverse-biased, and therefore not conduct any current. In fact, the only purpose for having D_1 in the circuit is to prevent transistor damage in the case of a *negative* voltage being impressed on the input (a voltage that is negative, rather than positive, with respect to ground).

With no voltage between the base and emitter of transistor Q_1, we would expect no current through it, either. However, as strange as it may seem, transistor Q_1 is not being used as is customary for a transistor.

In reality, Q_1 is being used in this circuit as nothing more than a back-to-back pair of diodes.

The following schematic shows the real function of Q_1:

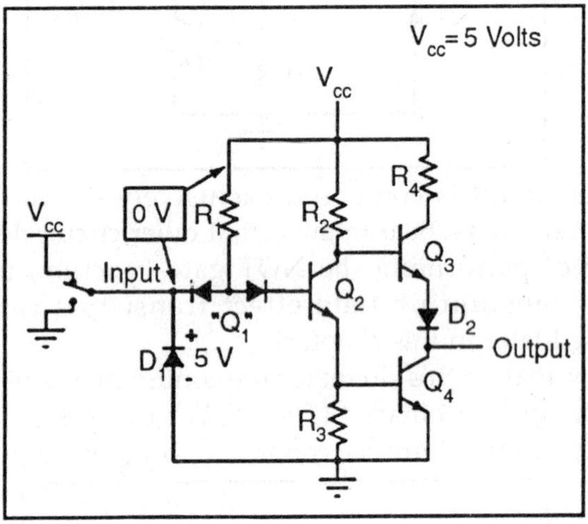

The purpose of these diodes is to "steer" current to or away from the base of transistor Q_2, depending on the logic level of the input.

Exactly how these two diodes are able to "steer" current isn't exactly obvious at first inspection, so a short example may be necessary for understanding.

Suppose we had the following diode/resistor circuit, representing the base-emitter junctions of transistors Q_2 and Q_4 as single diodes, stripping away all other portions of the circuit so that we can concentrate on the current "steered" through the two back-to-back diodes:

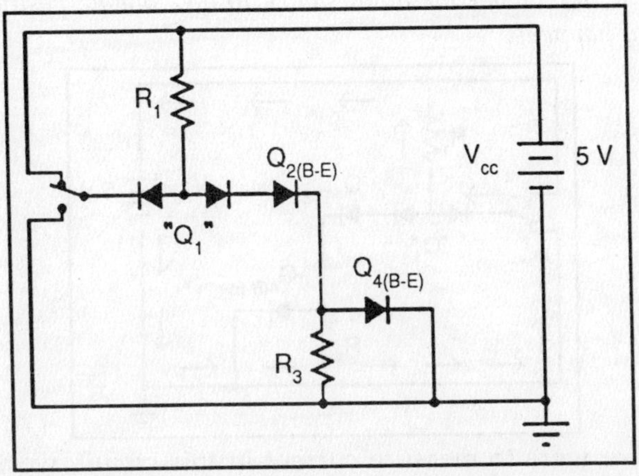

With the input switch in the "up" position (connected to V_{cc}), it should be obvious that there will be no current through the left steering diode of Q_1, because there isn't any voltage in the switch-diode-R_1-switch loop to motivate electrons to flow. However, there *will* be current through the right steering diode of Q_1, as well as through Q_2's base-emitter diode junction and Q_4's base-emitter diode junction:

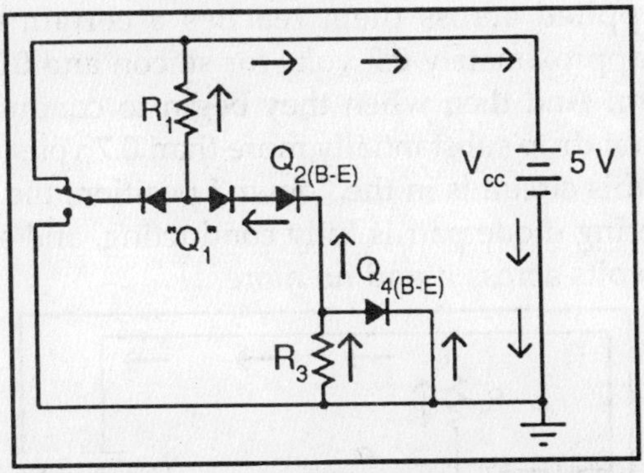

This tells us that in the real gate circuit, transistors Q_2 and Q_4 will have base current, which will turn them on to conduct collector current. The total voltage dropped between the base of Q_1 (the node joining the two back-to-back steering diodes) and ground will be about 2.1 volts, equal to the combined voltage drops of three PN junctions: the right steering diode, Q_2's base-emitter diode, and Q_4's base-emitter diode.

Now, let's move the input switch to the "down" position and see what happens:

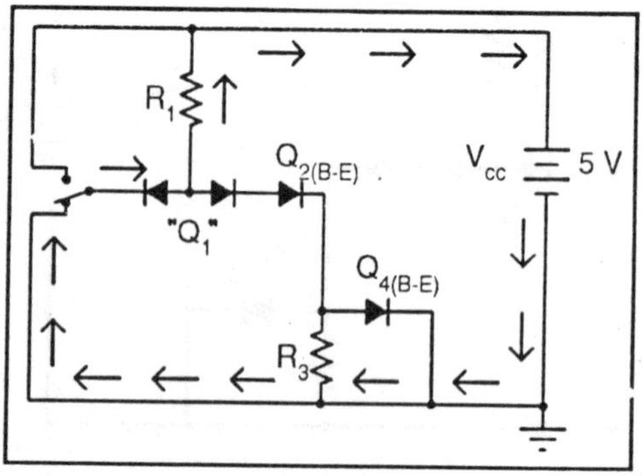

If we were to measure current in this circuit, we would find that *all* of the current goes through the left steering diode of $Q_1$and *none* of it through the right diode. Why is this? It still appears as though there is a complete path for current through Q_4's diode, Q_2's diode, the right diode of the pair, and R_1, so why will there be no current through that path?

Remember that PN junction diodes are very nonlinear devices: they do not even begin to conduct current until the forward voltage applied across them reaches a certain minimum quantity, approximately 0.7 volts for silicon and 0.3 volts for germanium. And then when they begin to conduct current, they will not drop substantially more than 0.7 volts. When the switch in this circuit is in the "down" position, the left diode of the steering diode pair is fully conducting, and so it drops about 0.7 volts across it and no more.

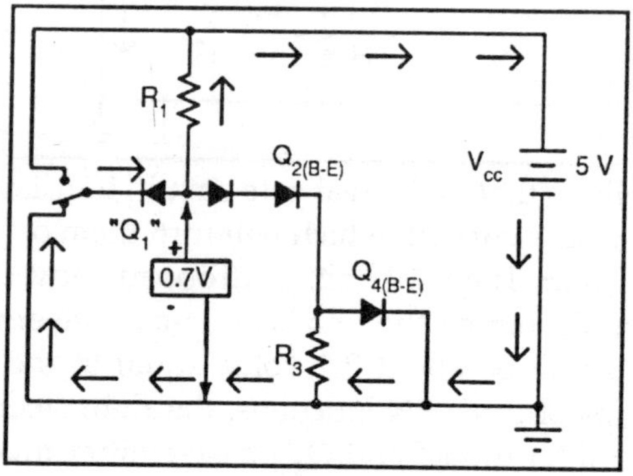

Recall that with the switch in the "up" position (transistors Q_2 and Q_4 conducting), there was about 2.1 volts dropped between those same two points (Q_1's base and ground), which also happens to be the *minimum* voltage necessary to forward-bias three series-connected silicon PN junctions into a state of conduction.

The 0.7 volts provided by the left diode's forward voltage drop is simply insufficient to allow any electron flow through the series string of the right diode, Q_2's diode, and the R_3// Q_4 diode parallel subcircuit, and so no electrons flow through that path. With no current through the bases of either transistor Q_2 or Q_4, neither one will be able to conduct collector current: transistors Q_2 and Q_4 will both be in a state of cutoff.

Consequently, this circuit configuration allows 100 per cent switching of Q_2 base current (and therefore control over the rest of the gate circuit, including voltage at the output) by diversion of current through the left steering diode.

In the case of our example gate circuit, the input is held "high" by the switch (connected to V_{cc}), making the left steering diode (zero voltage dropped across it). However, the right steering diode is conducting current through the base of Q_2, through resistor R_1:

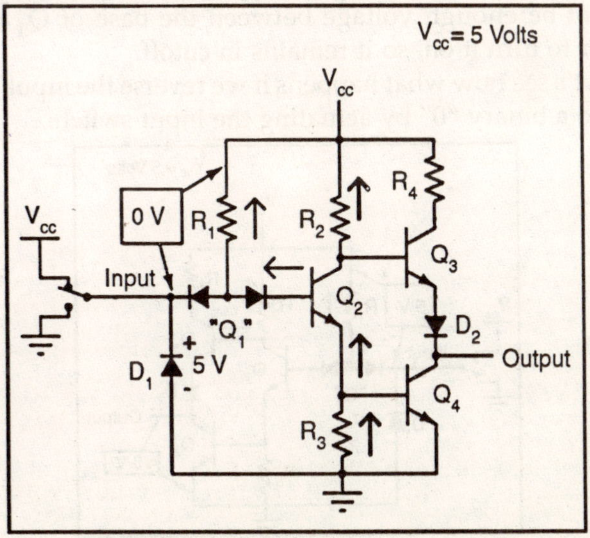

With base current provided, transistor Q_2 will be turned "on." More specifically, it will be *saturated* by virtue of the more-than-adequate current allowed by R_1 through the base. With Q_2 saturated, resistor R_3 will be dropping enough voltage to forward-bias the base-emitter junction of transistor Q_4, thus saturating it as well:

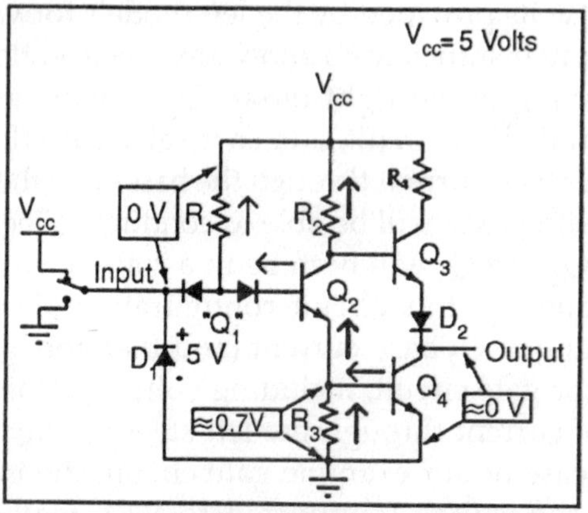

With Q_4 saturated, the output terminal will be almost directly shorted to ground, leaving the output terminal at a voltage (in reference to ground) of almost 0 volts, or a binary "0" ("low") logic level. Due to the presence of diode D_2, there will not be enough voltage between the base of Q_3 and its emitter to turn it on, so it remains in cutoff.

Let's see now what happens if we reverse the input's logic level to a binary "0" by actuating the input switch:

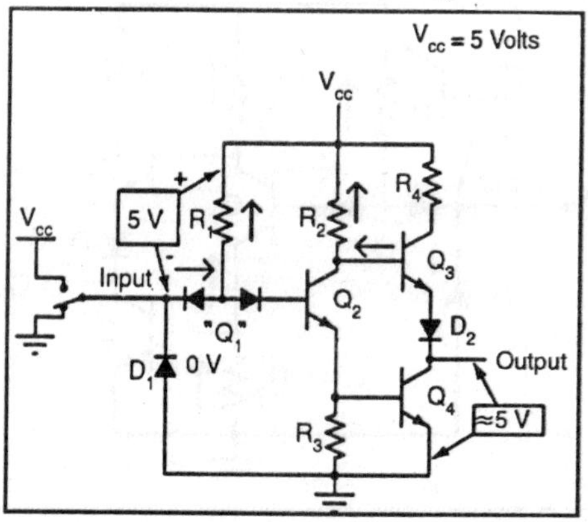

Now there will be current through the left steering diode of Q_1 and no current through the right steering diode. This eliminates current through the base of Q_2, thus turning it off. With Q_2 off, there is no longer a path for Q_4 base current, so Q_4 goes into cutoff as well.

Q_3, on the other hand, now has sufficient voltage dropped between its base and ground to forward-bias its base-emitter junction and saturate it, thus raising the output terminal voltage to a "high" state. In actuality, the output voltage will be somewhere around 4 volts depending on the degree of saturation and any load current, but still high enough to be considered a "high" (1) logic level.

With this, our simulation of the inverter circuit is complete: a "1" in gives a "0" out, and visa-versa. The astute observer will note that this inverter circuit's input will assume a "high" state of left floating (not connected to either V_{cc} or ground). With the input terminal left unconnected, there will be no current through the left steering diode of Q_1, leaving all of R_1's current to go through Q_2's base, thus saturating Q_2 and driving the circuit output to a "low" state:

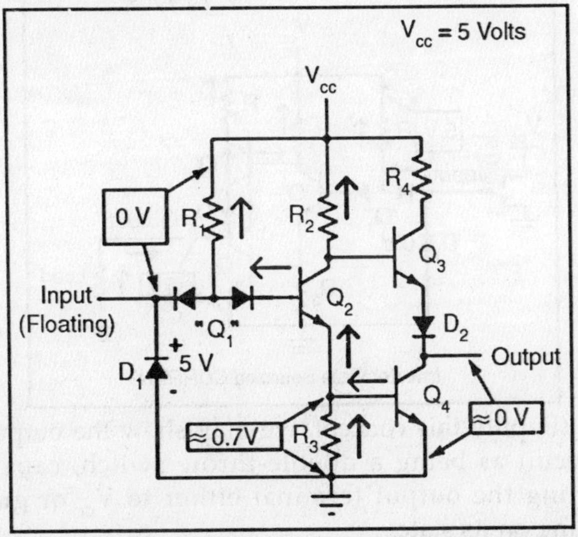

The tendency for such a circuit to assume a high input state if left floating is one shared by all gate circuits based on this type of design, known as Transistor-to-Transistor Logic, or *TTL*. This characteristic may be taken advantage of in simplifying the design of a gate's *output* circuitry, knowing that the outputs of gates typically drive the inputs of other gates. If the input of a TTL gate circuit assumes a high state when floating, then the output of any gate driving a TTL input need only provide a path to ground for a low state and be floating for a high state. This concept may require further elaboration for full understanding, so I will explore it in detail here.

A gate circuit as we have just analysed has the ability to handle output current in two directions: in and out. Technically, this is known as *sourcing* and *sinking* current, respectively.

When the gate output is high, there is continuity from the output terminal to V_{cc} through the top output transistor (Q_3), allowing electrons to flow from ground, through a load, into the gate's output terminal, through the emitter of Q_3, and eventually up to the V_{cc} power terminal (positive side of the DC power supply):

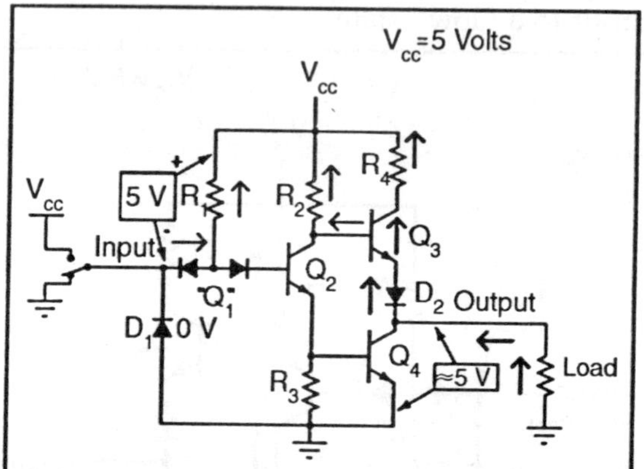

Inverter Gate Sourcing CURRENT

To simplify this concept, we may show the output of a gate circuit as being a double-throw switch, capable of connecting the output terminal either to V_{cc} or ground, depending on its state.

For a gate outputting a "high" logic level, the combination of Q_3 saturated and Q_4 cutoff is analogous to a double-throw switch in the "V_{cc}" position, providing a path for current through a grounded load:

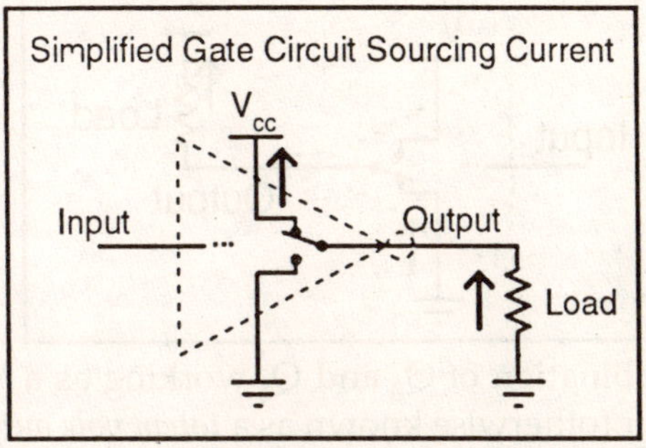

Please note that this two-position switch shown inside the gate symbol is representative of transistors Q_3 and Q_4alternately connecting the output terminal to V_{cc} or ground, *not* of the switch previously shown sending an input signal to the gate!

Conversely, when a gate circuit is outputting a "low" logic level to a load, it is analogous to the double-throw switch being set in the "ground" position. Current will then be going the other way if the load resistance connects to V_{cc}: from ground, through the emitter of Q_4, out the output terminal, through the load resistance, and back to V_{cc}. In this condition, the gate is said to be *sinking* current:

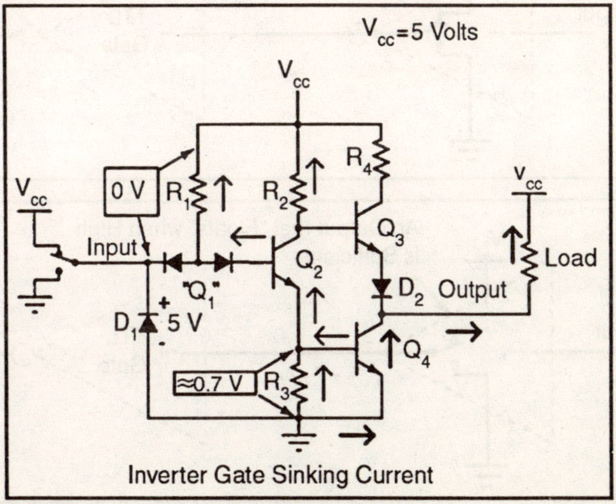

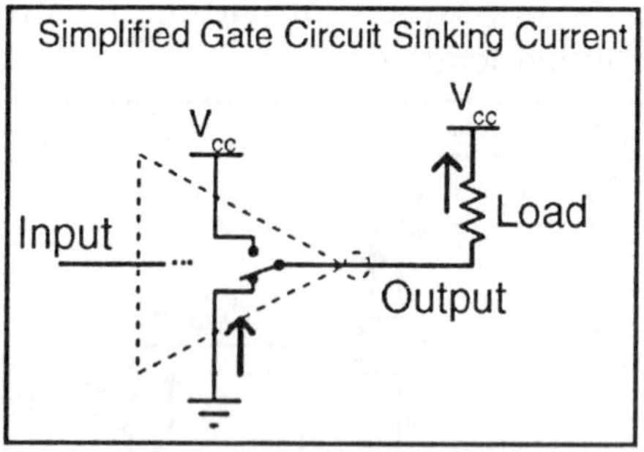

The combination of Q_3 and Q_4 working as a "push-pull" transistor pair (otherwise known as a *totem pole output*) has the ability to either source current (draw in current to V_{cc}) or sink current (output current from ground) to a load. However, a standard TTL gate *input* never needs current to be sourced, only sunk. That is, since a TTL gate input naturally assumes a high state if left floating, any gate output driving a TTL input need only sink current to provide a "0" or "low" input, and need not source current to provide a "1" or a "high" logic level at the input of the receiving gate:

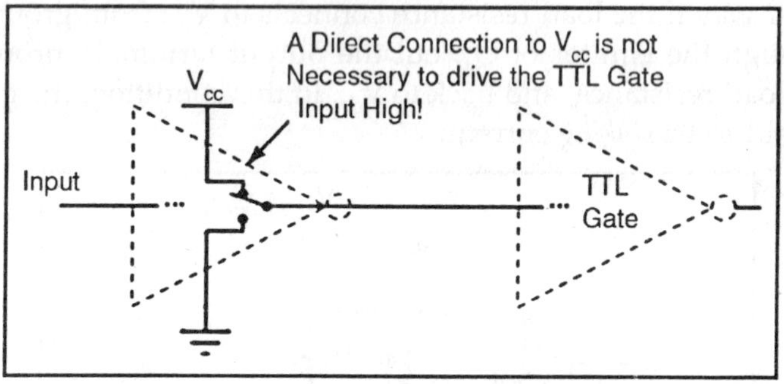

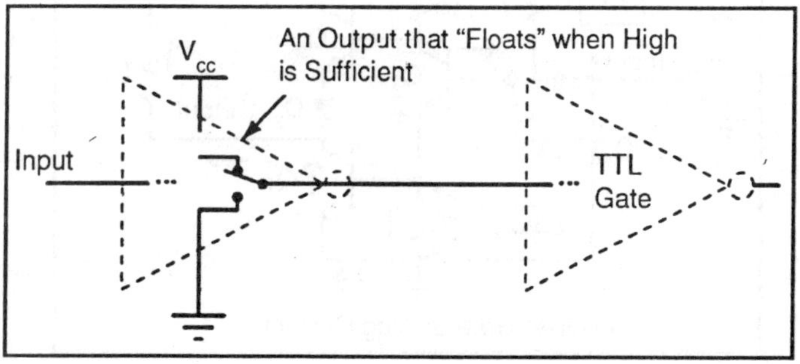

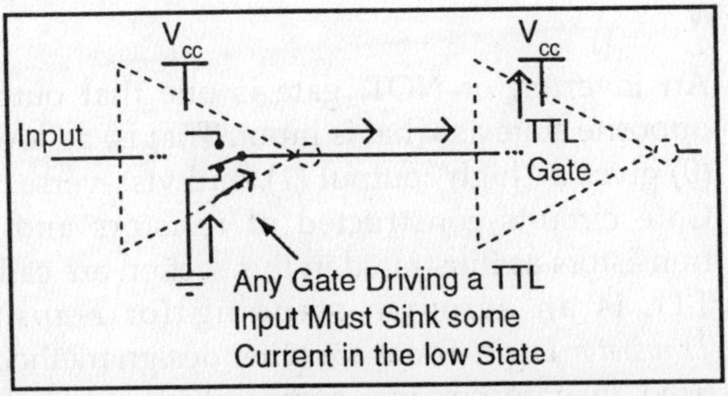

This means we have the option of simplifying the output stage of a gate circuit so as to eliminate Q_3 altogether. *The result is known as an open-collector output*:

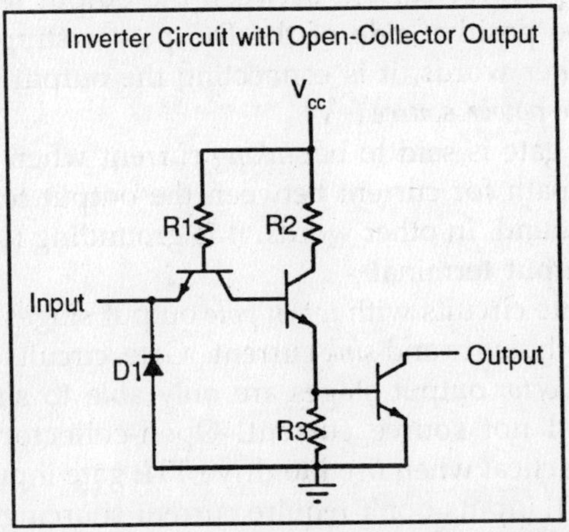

To designate open-collector output circuitry within a standard gate symbol, a special marker is used. Shown here is the symbol for an inverter gate with open-collector output:

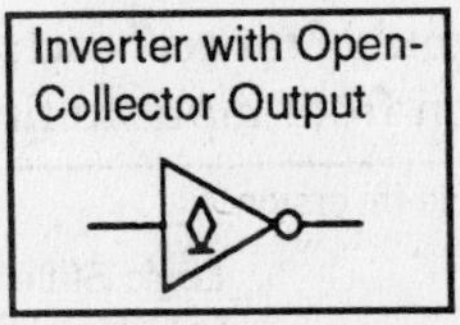

Please keep in mind that the "high" default condition of a floating gate input is only true for TTL circuitry, and not necessarily for other types, especially for logic gates constructed of field-effect transistors.

REVIEW

- An inverter, or NOT, gate is one that outputs the opposite state as what is input. That is, a "low" input (0) gives a "high" output (1), and visa-versa.
- Gate circuits constructed of resistors and bipolar transistors as illustrated in this section are called *TTL*. TTL is an acronym standing for *Transistor-to-Transistor Logic*. There are other design methodologies used in gate circuits, some which use field-effect transistors rather than bipolar transistors.
- A gate is said to be *sourcing* current when it provides a path for current between the output terminal and the positive side of the DC power supply (V_{cc}). In other words, it is connecting the output terminal to the *power source* (+V).
- A gate is said to be *sinking* current when it provides a path for current between the output terminal and ground. In other words, it is grounding (sinking) the output terminal.
- Gate circuits with *totem pole* output stages are able to both *source* and *sink* current. Gate circuits with *open-collector* output stages are only able to sink current, and not source current. Open-collector gates are practical when used to drive TTL gate inputs because TTL inputs don't require current sourcing.

THE "BUFFER" GATE

If we were to connect two inverter gates together so that the output of one fed into the input of another, the two inversion functions would "cancel" each other out so that there would be no inversion from input to final output:

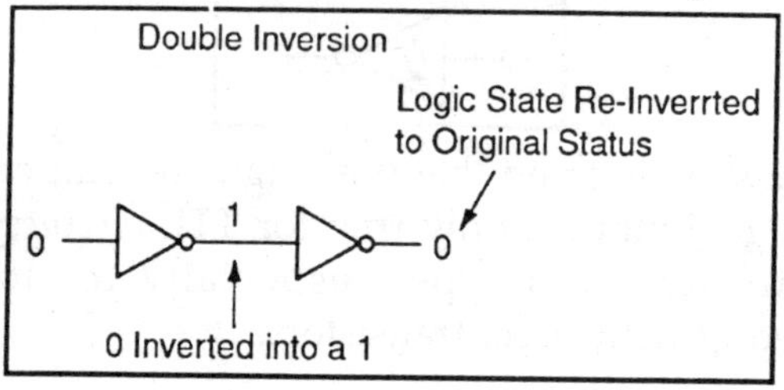

While this may seem like a pointless thing to do, it does have practical application. Remember that gate circuits are signal*amplifiers*, regardless of what logic function they may perform. A weak signal source (one that is not capable of sourcing or sinking very much current to a load) may be boosted by means of two inverters like the pair shown in the previous illustration. The logic level is unchanged, but the full current-sourcing or -sinking capabilities of the final inverter are available to drive a load resistance if needed.

For this purpose, a special logic gate called a *buffer* is manufactured to perform the same function as two inverters. *Its symbol is simply a triangle, with no inverting "bubble" on the output terminal*:

"Buffer" Gate

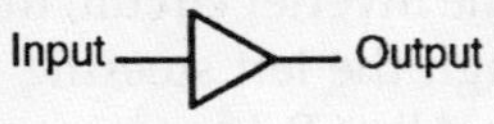

Input	Output
0	0
1	1

The internal schematic diagram for a typical open-collector buffer is not much different from that of a simple inverter: only one more common-emitter transistor stage is added to re-invert the output signal.

Buffer Circuit with Open- Collector Output

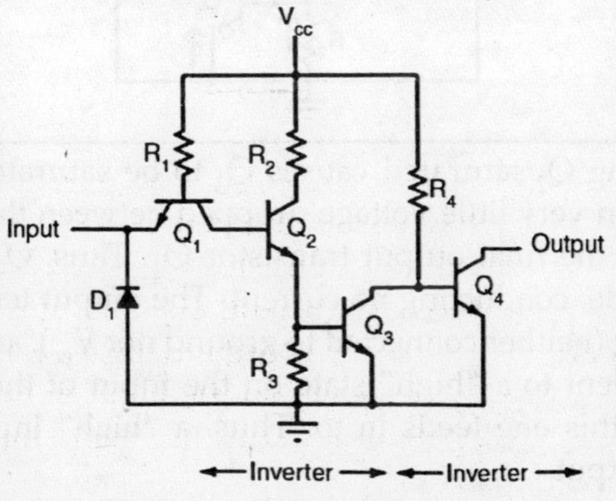

Let's analyse this circuit for two conditions: an input logic level of "1" and an input logic level of "0." First, a "high" (1) input:

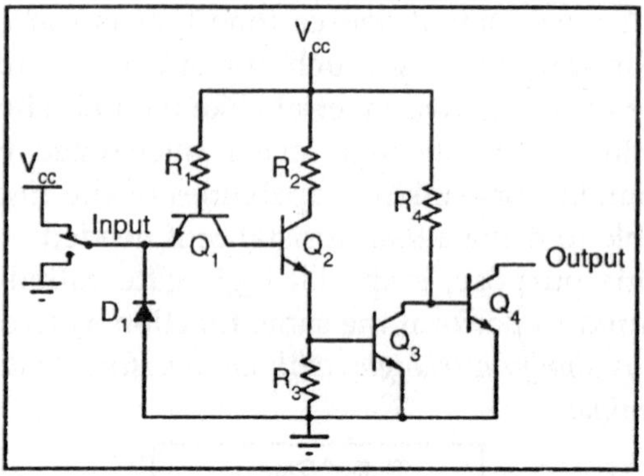

As before with the inverter circuit, the "high" input causes no conduction through the left steering diode of Q_1 (emitter-to-base PN junction). All of R_1's current goes through the base of transistor Q_2, saturating it:

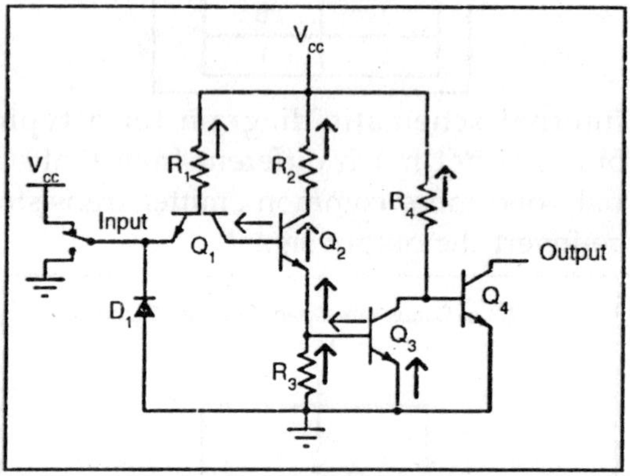

Having Q_2 saturated causes Q_3 to be saturated as well, resulting in very little voltage dropped between the base and emitter of the final output transistor Q_4. Thus, Q_4 will be in cutoff mode, conducting no current. The output terminal will be floating (neither connected to ground nor V_{cc}), and this will be equivalent to a "high" state on the input of the next TTL gate that this one feeds in to. Thus, a "high" input gives a "high" output.

With a "low" input signal (input terminal grounded), the analysis looks something like this:

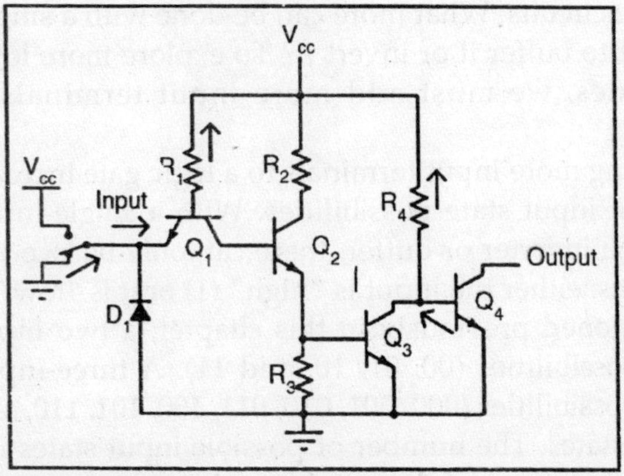

All of R_1's current is now diverted through the input switch, thus eliminating base current through Q_2. This forces transistor Q_2 into cutoff so that no base current goes through Q_3 either. With Q_3 cutoff as well, Q_4 is will be saturated by the current through resistor R_4, thus connecting the output terminal to ground, making it a "low" logic level. Thus, a "low" input gives a "low" output.

The schematic diagram for a buffer circuit with totem pole output transistors is a bit more complex, but the basic principles, and certainly the truth table, are the same as for the open-collector circuit:

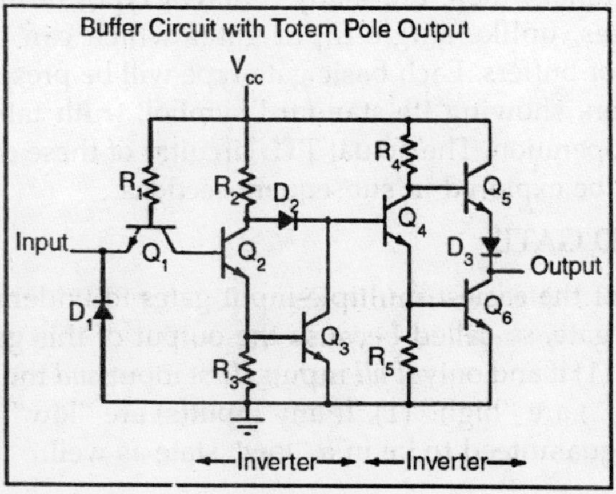

MULTIPLE-INPUT GATES

Inverters and buffers exhaust the possibilities for single-input gate circuits. What more can be done with a single logic signal but to buffer it or invert it? To explore more logic gate possibilities, we must add more input terminals to the circuit(s).

Adding more input terminals to a logic gate increases the number of input state possibilities. With a single-input gate such as the inverter or buffer, there can only be two possible input states: either the input is "high" (1) or it is "low" (0). As was mentioned previously in this chapter, a two input gate has *four* possibilities (00, 01, 10, and 11). A three-input gate has*eight* possibilities (000, 001, 010, 011, 100, 101, 110, and 111) for input states. The number of possible input states is equal to two to the power of the number of inputs:

Number of Possible Input States=2^n

Where:

n=Number of Inputs

This increase in the number of possible input states obviously allows for more complex gate behaviour. Now, instead of merely inverting or amplifying (buffering) a single "high" or "low" logic level, the output of the gate will be determined by whatever *combination* of 1's and 0's is present at the input terminals.

Since so many combinations are possible with just a few input terminals, there are many different types of multiple-input gates, unlike single-input gates which can only be inverters or buffers. Each basic gate type will be presented in this section, showing its standard symbol, truth table, and practical operation. The actual TTL circuitry of these different gates will be explored in subsequent sections.

THE AND GATE

One of the easiest multiple-input gates to understand is the AND gate, so-called because the output of this gate will be "high" (1) if and only if *all* inputs (first input *and* the second input *and* ...) are "high" (1). If any input(s) are "low" (0), the output is guaranteed to be in a "low" state as well.

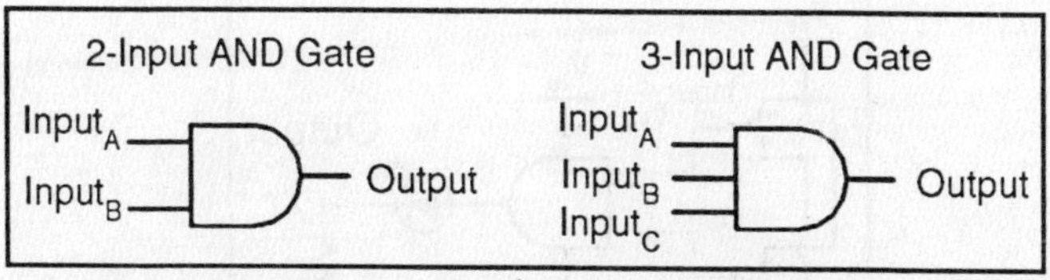

In case you might have been wondering, AND gates are made with more than three inputs, but this is less common than the simple two-input variety.

A two-input and Gate's Truth Table Looks Like this:

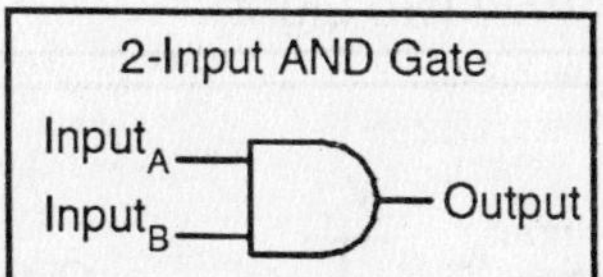

A	B	Output
0	0	0
0	1	0
1	0	0
1	1	1

What this truth table means in practical terms is shown in the following sequence of illustrations, with the 2-input AND gate subjected to all possibilities of input logic levels. An LED (Light-Emitting Diode) provides visual indication of the output logic level:

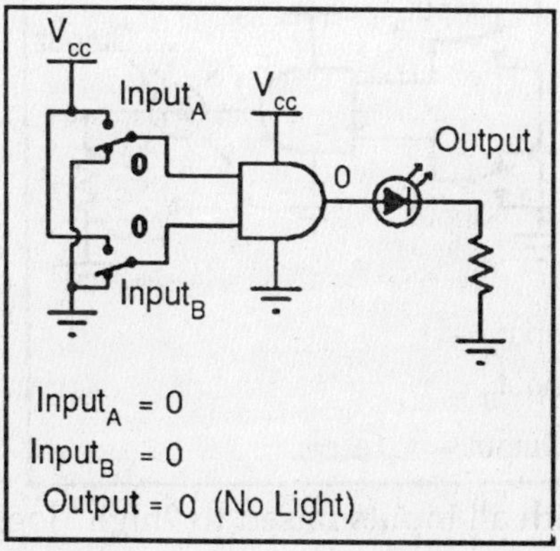

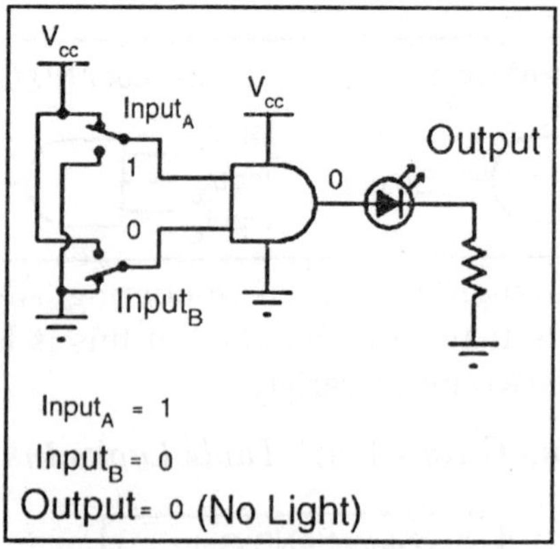

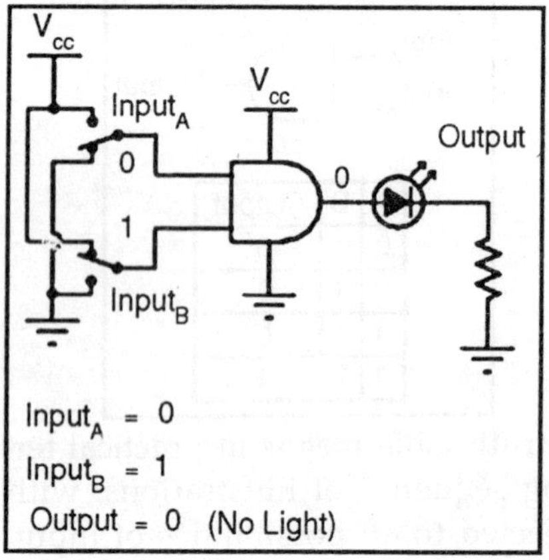

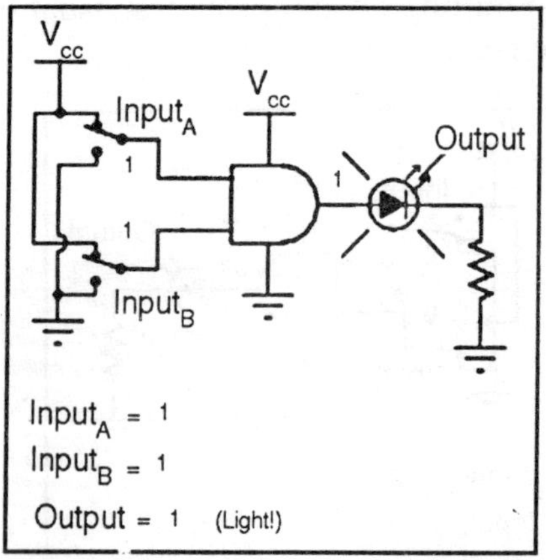

It is only with all inputs raised to "high" logic levels that the AND gate's output goes "high," thus energizing the LED for only one out of the four input combination states.

THE NAND GATE

A variation on the idea of the AND gate is called the NAND gate. The word "NAND" is a verbal contraction of the words NOT and AND. Essentially, a NAND gate behaves the same as an AND gate with a NOT (inverter) gate connected to the output terminal. To symbolize this output signal inversion, the NAND gate symbol has a bubble on the output line. The truth table for a NAND gate is as one might expect, exactly opposite as that of an AND gate:

2-Input NAND Gate

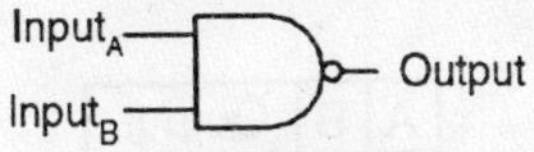

A	B	Output
0	0	1
0	1	1
1	0	1
1	1	0

Equivalent Gate Circuit

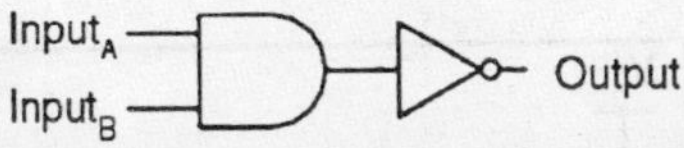

As with AND gates, NAND gates are made with more than two inputs. In such cases, the same general principle applies: the output will be "low" (0) if and only if all inputs are "high" (1). If any input is "low" (0), the output will go "high" (1).

THE OR GATE

Our next gate to investigate is the OR gate, so-called because the output of this gate will be "high" (1) if *any* of the inputs (first input *or* the second input *or* ...) are "high" (1). The output of an OR gate goes "low" (0) if and only if all inputs are "low" (0).

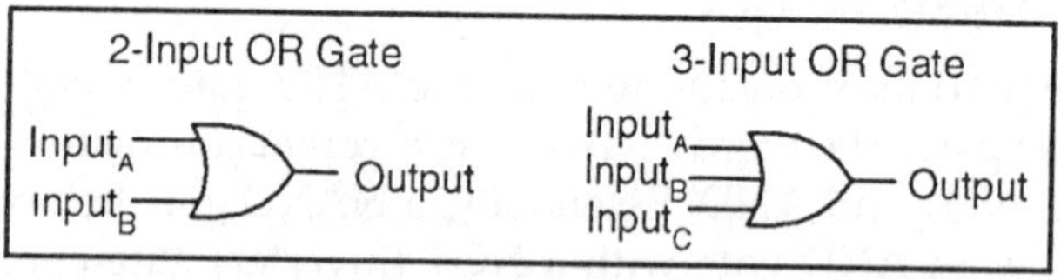

A two-input OR gate's truth table looks like this:

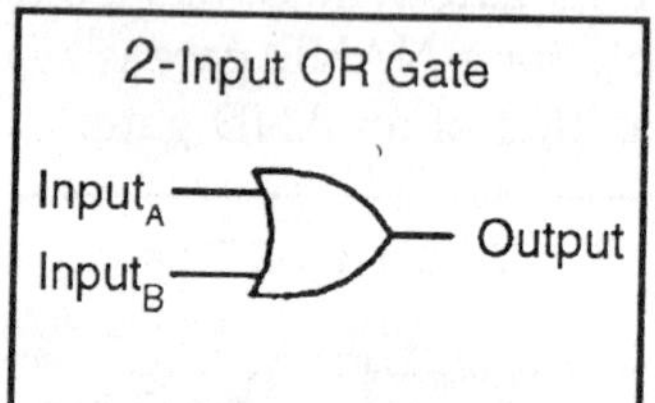

A	B	Output
0	0	0
0	1	1
1	0	1
1	1	1

The following sequence of illustrations demonstrates the OR gate's function, with the 2-inputs experiencing all possible logic levels. An LED (Light-Emitting Diode) provides visual indication of the gate's output logic level:

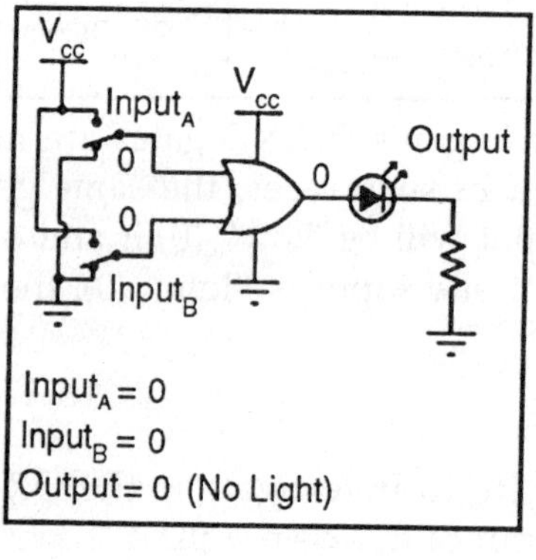

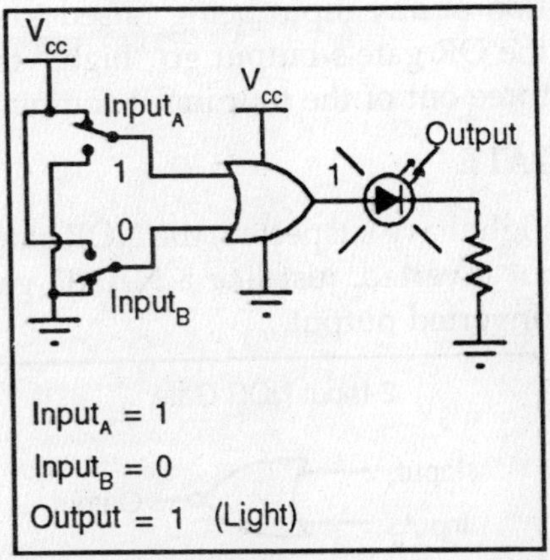

V_{cc}
Input_A
V_{cc}
Output
1
1
0
Input_B
$\text{Input}_A = 1$
$\text{Input}_B = 0$
Output = 1 (Light)

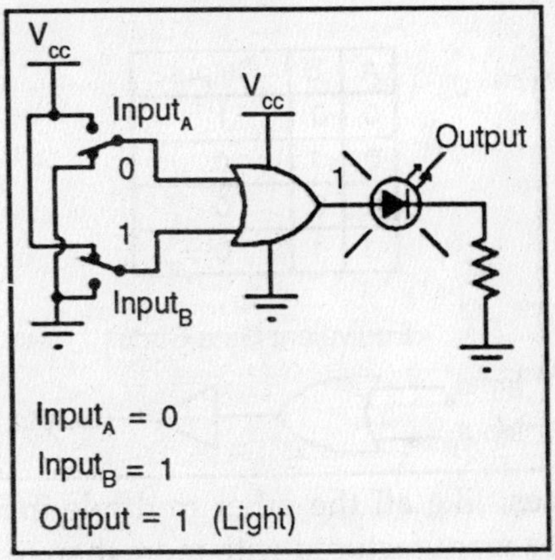

V_{cc}
Input_A
V_{cc}
Output
0
1
1
Input_B
$\text{Input}_A = 0$
$\text{Input}_B = 1$
Output = 1 (Light)

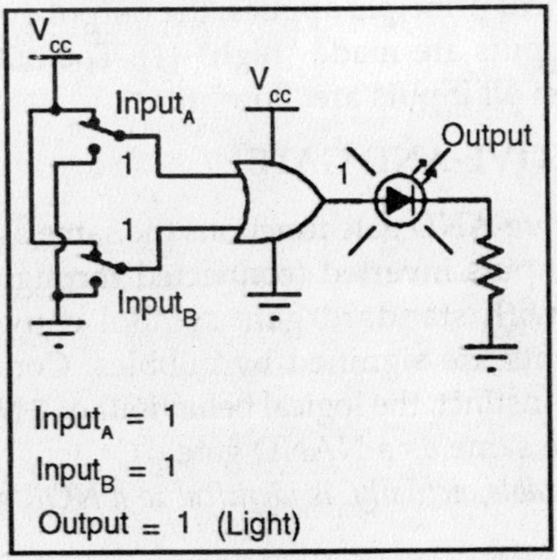

V_{cc}
Input_A
V_{cc}
Output
1
1
1
Input_B
$\text{Input}_A = 1$
$\text{Input}_B = 1$
Output = 1 (Light)

A condition of any input being raised to a "high" logic level makes the OR gate's output go "high," thus energizing the LED for three out of the four input combination states.

THE NOR GATE

As you might have suspected, the NOR gate is an OR gate with its output inverted, just like a NAND gate is an AND gate with an inverted output.

2-Input NOR Gate

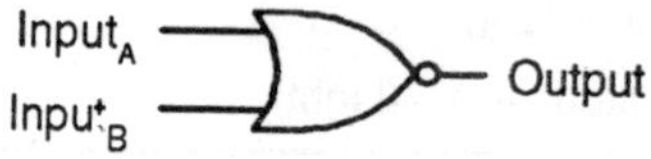

A	B	Output
0	0	1
0	1	0
1	0	0
1	1	0

Equivalent Gate Circuit

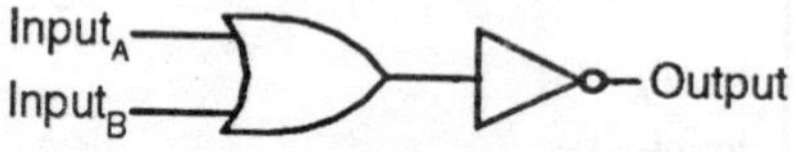

NOR gates, like all the other multiple-input gates seen thus far, can be manufactured with more than two inputs. Still, the same logical principle applies: the output goes "low" (0) if any of the inputs are made "high" (1). The output is "high" (1) only when all inputs are "low" (0).

THE NEGATIVE-AND GATE

A Negative-AND gate functions the same as an AND gate with all its inputs inverted (connected through NOT gates). In keeping with standard gate symbol convention, these inverted inputs are signified by bubbles. Contrary to most peoples' first instinct, the logical behaviour of a Negative-AND gate is *not* the same as a NAND gate.

Its truth table, actually, is identical to a NOR gate:

2- Input Negative-And Gate

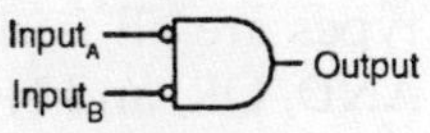

A	B	Output
0	0	1
0	1	0
1	0	0
1	1	0

Equivalent Gate Circuits

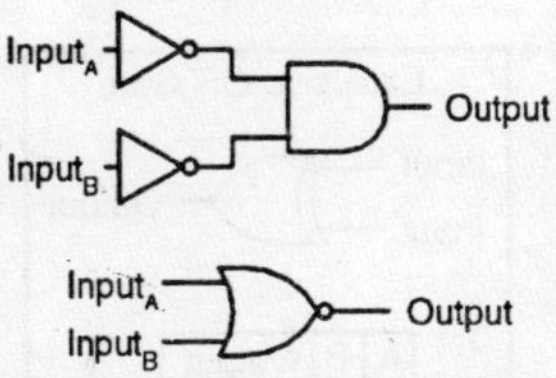

THE NEGATIVE-OR GATE

Following the same pattern, a Negative-OR gate functions the same as an OR gate with all its inputs inverted. In keeping with standard gate symbol convention, these inverted inputs are signified by bubbles. The behaviour and truth table of a Negative-OR gate is the same as for a NAND gate:

2-Input Negative-OR Gate

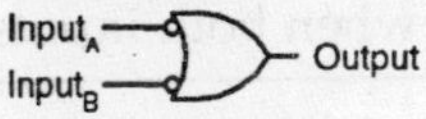

A	B	Output
0	0	1
0	1	1
1	0	1
1	1	0

Equivalent Gate Circuits

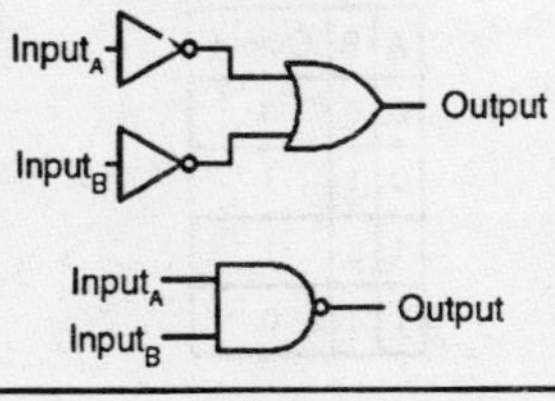

THE EXCLUSIVE-OR GATE

The last six gate types are all fairly direct variations on three basic functions: AND, OR, and NOT. The Exclusive-OR gate, however, is something quite different.

Exclusive-OR gates output a "high" (1) logic level if the inputs are at *different* logic levels, either 0 and 1 or 1 and 0. Conversely, they output a "low" (0) logic level if the inputs are at the *same* logic levels. The Exclusive-OR (sometimes called XOR) gate has both a symbol and a truth table pattern that is unique:

Exclusive-OR Gate

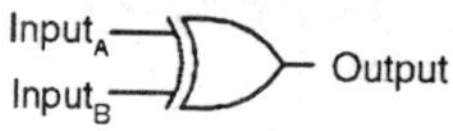

A	B	Output
0	0	0
0	1	1
1	0	1
1	1	0

There are equivalent circuits for an Exclusive-OR gate made up of AND, OR, and NOT gates, just as there were for NAND, NOR, and the negative-input gates. A rather direct approach to simulating an Exclusive-OR gate is to start with a regular OR gate, then add additional gates to inhibit the output from going "high" (1) when both inputs are "high" (1):

Exclusive OR Equivalent Circuit

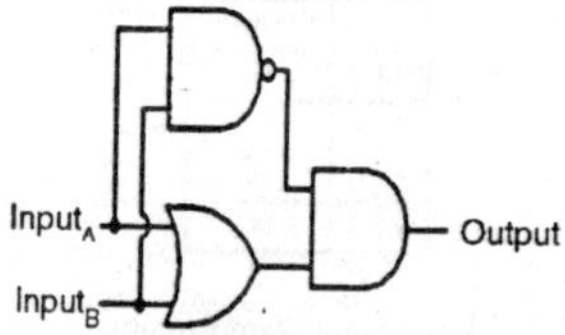

A	B	Output
0	0	0
0	1	1
1	0	1
1	1	0

In this circuit, the final AND gate acts as a buffer for the output of the OR gate whenever the NAND gate's output is high, which it is for the first three input state combinations (00, 01, and 10).

However, when both inputs are "high" (1), the NAND gate outputs a "low" (0) logic level, which forces the final AND gate to produce a "low" (0) output.

Another equivalent circuit for the Exclusive-OR gate uses a strategy of two AND gates with inverters, set up to generate "high" (1) outputs for input conditions 01 and 10. A final OR gate then allows either of the AND gates' "high" outputs to create a final "high" output:

Exclusive-OR Equivalent Circuit

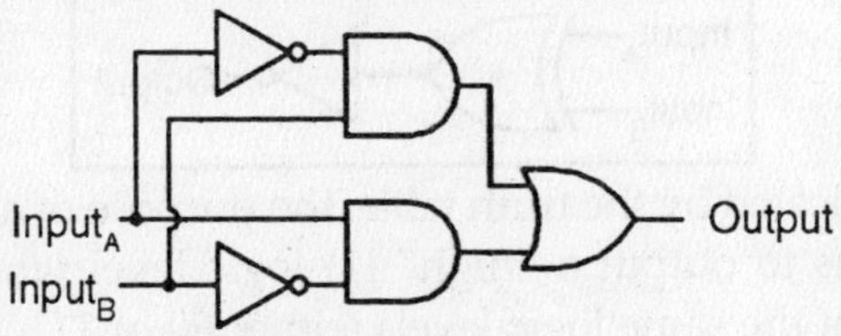

A	B	Output
0	0	0
0	1	1
1	0	1
1	1	0

Exclusive-OR gates are very useful for circuits where two or more binary numbers are to be compared bit-for-bit, and also for error detection (parity check) and code conversion (binary to Grey and visa-versa).

THE EXCLUSIVE-NOR GATE

Finally, our last gate for analysis is the Exclusive-NOR gate, otherwise known as the XNOR gate.

It is equivalent to an Exclusive-OR gate with an inverted output.

The truth table for this gate is exactly opposite as for the Exclusive-OR gate:

Exclusive-NOR Gate

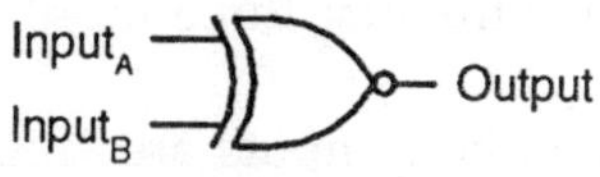

A	B	Output
0	0	1
0	1	0
1	0	0
1	1	1

Equivalent Gate Circuits

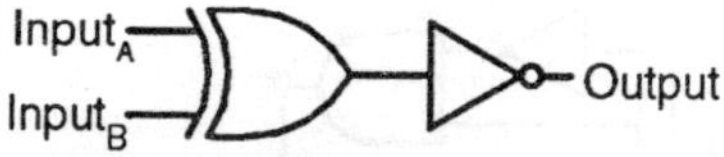

As indicated by the truth table, the purpose of an Exclusive-NOR gate is to output a "high" (1) logic level whenever both inputs are at the same logic levels (either 00 or 11).

TTL NAND GATES

Suppose we altered our basic open-collector inverter circuit, adding a second input terminal just like the first:

A Two-Input Inverter Circuit

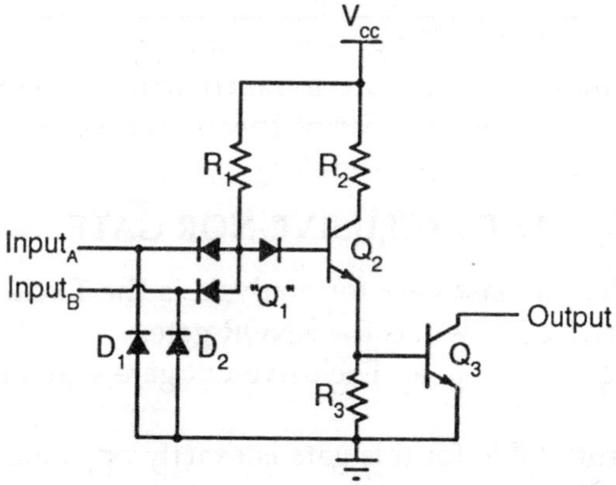

This schematic illustrates a real circuit, but it isn't called a "two-input inverter." Through analysis we will discover what this circuit's logic function is and correspondingly what it should be designated as.

Just as in the case of the inverter and buffer, the "steering" diode cluster marked "Q_1" is actually formed like a transistor, even though it isn't used in any amplifying capacity. Unfortunately, a simple NPN transistor structure is inadequate to simulate the *three* PN junctions necessary in this diode network, so a different transistor (and symbol) is needed. This transistor has one collector, one base, and *two* emitters, and in the circuit it looks like this:

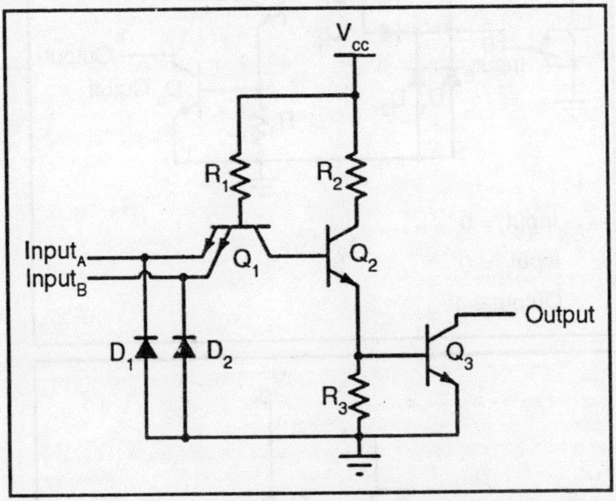

In the single-input (inverter) circuit, grounding the input resulted in an output that assumed the "high" (1) state. In the case of the open-collector output configuration, this "high" state was simply "floating." Allowing the input to float (or be connected to V_{cc}) resulted in the output becoming grounded, which is the "low" or 0 state. Thus, a 1 in resulted in a 0 out, and visa-versa.

Since this circuit bears so much resemblance to the simple inverter circuit, the only difference being a second input terminal connected in the same way to the base of transistor Q_2, we can say that each of the inputs will have the same effect on the output. Namely, if either of the inputs are grounded, transistor Q_2 will be forced into a condition of cutoff, thus turning Q_3 off and floating the output (output goes "high"). The following series of illustrations shows this for three input states (00, 01, and 10):

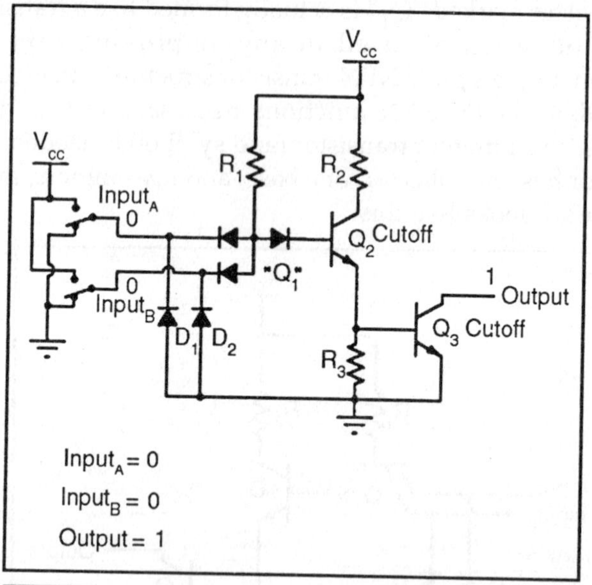

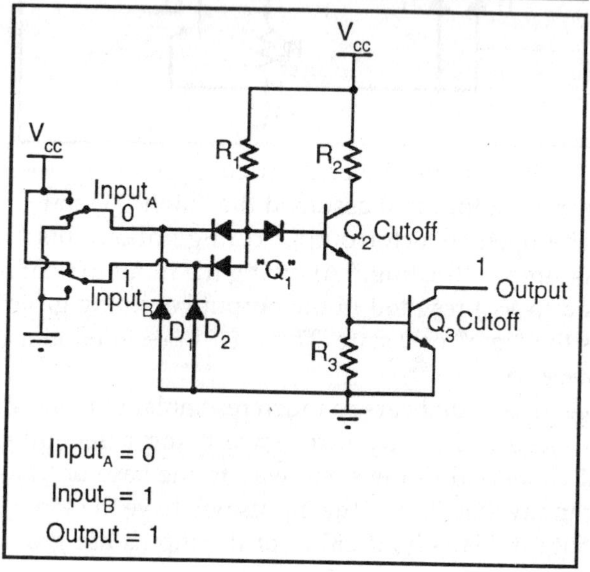

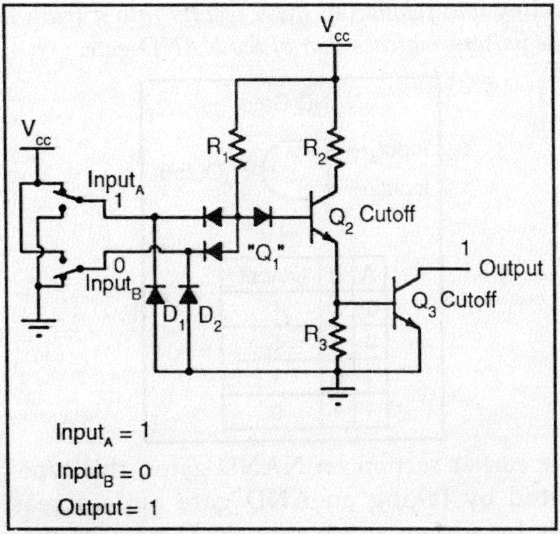

In any case where there is a grounded ("low") input, the output is guaranteed to be floating ("high"). Conversely, the only time the output will ever go "low" is if transistor Q_3 turns on, which means transistor Q_2 must be turned on (saturated), which means neither input can be diverting R_1 current away from the base of Q_2. The only condition that will satisfy this requirement is when both inputs are "high" (1):

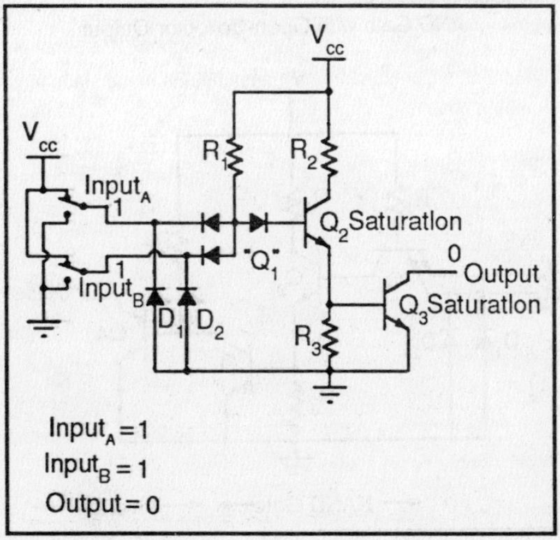

Collecting and tabulating these results into a truth table, we see that the pattern matches that of the NAND gate:

NAND Gate

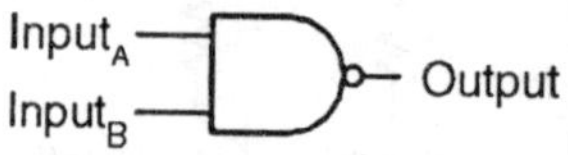

A	B	Output
0	0	1
0	1	1
1	0	1
1	1	0

In the earlier section on NAND gates, this type of gate was created by taking an AND gate and increasing its complexity by adding an inverter (NOT gate) to the output. However, when we examine this circuit, we see that the NAND function is actually the simplest, most natural mode of operation for this TTL design. To create an AND function using TTL circuitry, we need to *increase* the complexity of this circuit by adding an inverter stage to the output, just like we had to add an additional transistor stage to the TTL inverter circuit to turn it into a buffer:

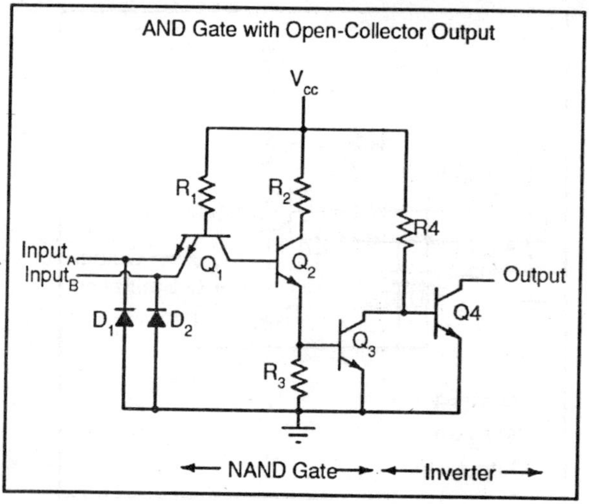

The truth table and equivalent gate circuit (an inverted-output NAND gate) are shown here:

AND Gate

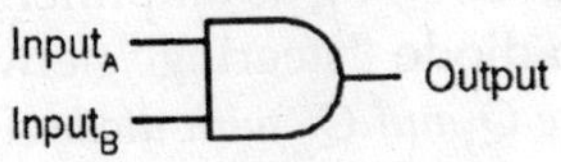

A	B	Output
0	0	0
0	1	0
1	0	0
1	1	1

Equivalent Circuit

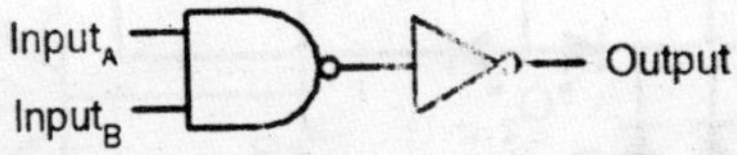

Of course, both NAND and AND gate circuits may be designed with totem-pole output stages rather than open-collector. I am opting to show the open-collector versions for the sake of simplicity.

TTL NOR AND OR GATES

Let's examine the following TTL circuit and analyse its operation:

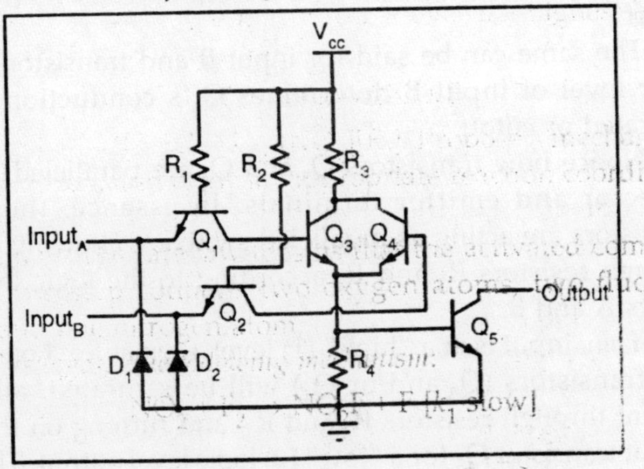

Transistors Q_1 and Q_2 are both arranged in the same manner that we've seen for transistor Q_1 in all the other TTL circuits.

Rather than functioning as amplifiers, Q_1 and Q_2 are both being used as two-diode "steering" networks.

We may replace Q_1 and Q_2 with diode sets to help illustrate:

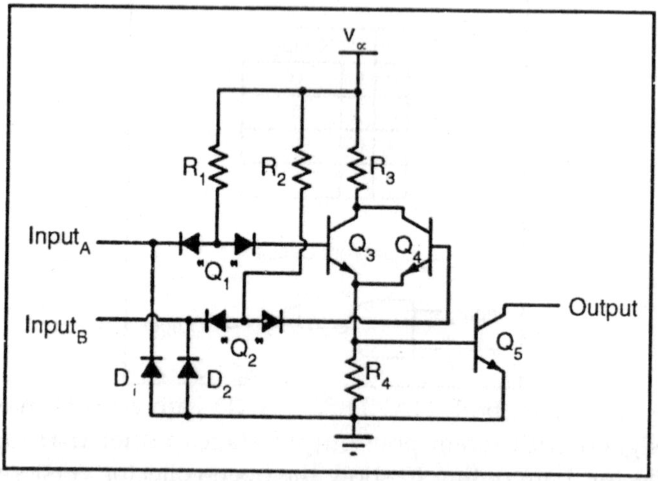

If input A is left floating (or connected to V_{cc}), current will go through the base of transistor Q_3, saturating it. If input A is grounded, that current is diverted away from Q_3's base through the left steering diode of "Q_1," thus forcing Q_3 into cutoff.

The same can be said for input B and transistor Q_4: the logic level of input B determines Q_4's conduction: either saturated or cutoff.

Notice how transistors Q_3 and Q_4 are paralleled at their collector and emitter terminals. In essence, these two transistors are acting as paralleled switches, allowing current through resistors R_3 and R_4 according to the logic levels of inputs A and B.

If *any* input is at a "high" (1) level, then at least one of the two transistors (Q_3 and/or Q_4) will be saturated, allowing current through resistors R_3 and R_4, and turning on the final output transistor Q_5 for a "low" (0) logic level output. The only way the output of this circuit can ever assume a "high" (1) state is if *both* Q_3 and Q_4 are cutoff, which means *both* inputs would have to be grounded, or "low" (0).

This circuit's truth table, then, is equivalent to that of the NOR gate:

NOR Gate

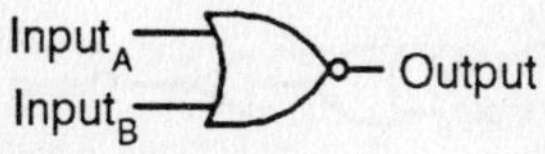

A	B	Output
0	0	1
0	1	0
1	0	0
1	1	0

In order to turn this NOR gate circuit into an OR gate, we would have to invert the output logic level with another transistor stage, just like we did with the NAND-to-AND gate example:

OR Gate with Open-Collector Output

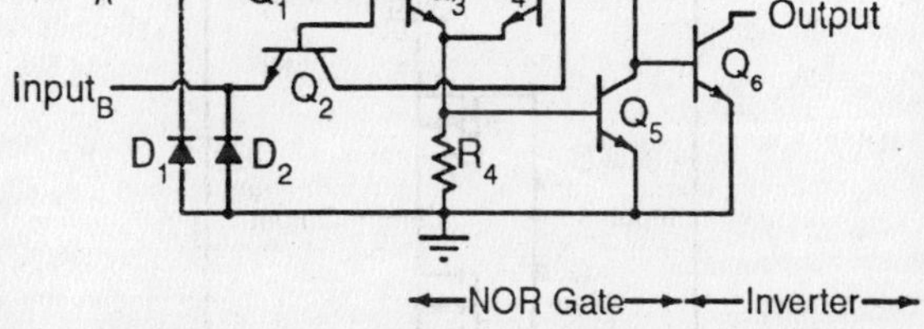

The truth table and equivalent gate circuit (an inverted-output NOR gate) are shown here:

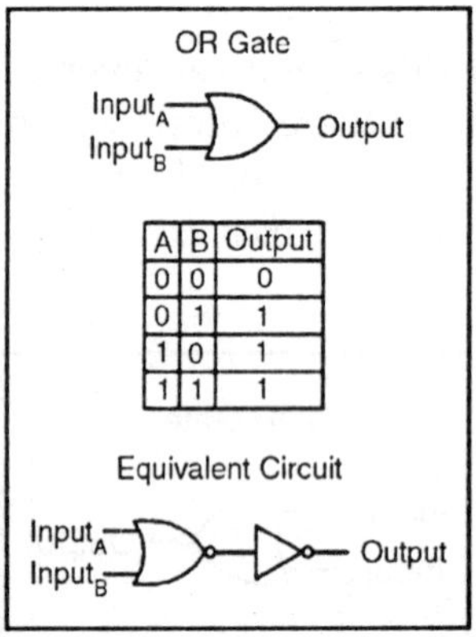

Of course, totem-pole output stages are also possible in both NOR and OR TTL logic circuits.

CMOS GATE

Up until this point, our analysis of transistor logic circuits has been limited to the *TTL* design paradigm, whereby bipolar transistors are used, and the general strategy of floating inputs being equivalent to "high" (connected to V_{cc}) inputs — and correspondingly, the allowance of "open-collector" output stages — is maintained. This, however, is not the only way we can build logic gates. Field-effect transistors, particularly the insulated-gate variety, may be used in the design of gate circuits. Being voltage-controlled rather than current-controlled devices, IGFETs tend to allow very simple circuit designs. Take for instance, the following inverter circuit built using P- and N-channel IGFETs:

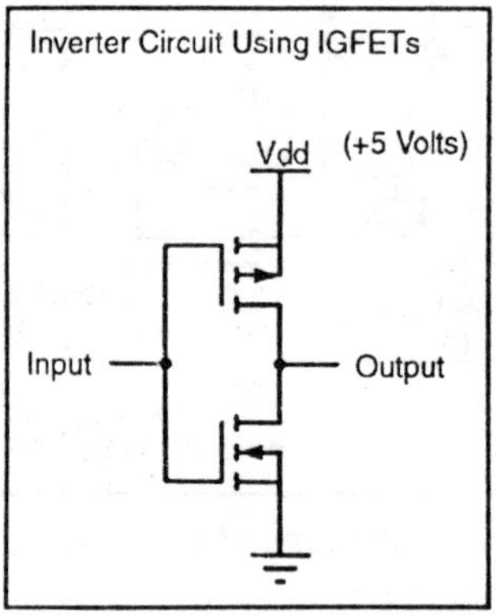

Notice the "V_{dd}" label on the positive power supply terminal. This label follows the same convention as "V_{cc}" in TTL circuits: it stands for the constant voltage applied to the drain of a field effect transistor, in reference to ground.

Let's connect this gate circuit to a power source and input switch, and examine its operation. Please note that these IGFET transistors are E-type and so are *normally-off* devices. It takes an applied voltage between gate and drain of the correct polarity to bias them *on*.

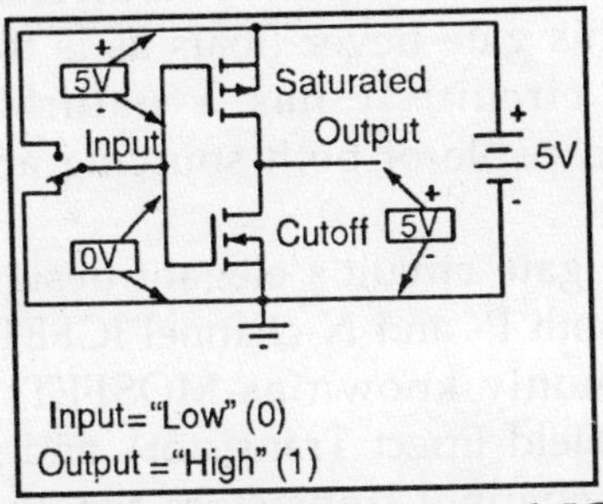

The upper transistor is a P-channel IGFET. When the channel the channel is enhanced and current is allowed between source and drain. So, in the above illustration, the top transistor is turned on.

The lower transistor, having zero voltage between gate and substrate (source), is in its normal mode: *off*. Thus, the action of these two transistors are such that the output terminal of the gate circuit has a solid connection to V_{dd} and a very high resistance connection to ground. This makes the output "high" (1) for the "low" (0) state of the input. *Next, we'll move the input switch to its other position and see what happens*:

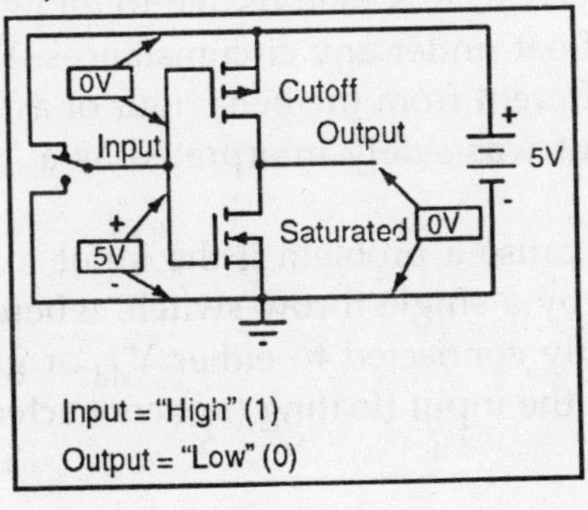

Now the lower transistor (N-channel) is saturated because it has sufficient voltage of the correct polarity applied between gate and substrate (channel) to turn it on (positive on gate, negative on the channel).

The upper transistor, having zero voltage applied between its gate and substrate, is in its normal mode: *off*. Thus, the output of this gate circuit is now "low" (0). Clearly, this circuit exhibits the behaviour of an inverter, or NOT gate.

Using field-effect transistors instead of bipolar transistors has greatly simplified the design of the inverter gate. Note that the output of this gate never floats as is the case with the simplest TTL circuit: it has a natural "totem-pole" configuration, capable of both sourcing and sinking load current.

Key to this gate circuit's elegant design is the *complementary* use of both P- and N-channel IGFETs. Since IGFETs are more commonly known as MOSFETs (Metal-Oxide-Semiconductor Field Effect Transistor), and this circuit uses both P- and N-channel transistors together, the general classification given to gate circuits like this one is *CMOS*: Complementary Metal Oxide Semiconductor.

CMOS circuits aren't plagued by the inherent nonlinearities of the field-effect transistors, because as digital circuits their transistors always operate in either the *saturated* or *cutoff* modes and never in the *active* mode.

Their inputs are, however, sensitive to high voltages generated by electrostatic (static electricity) sources, and may even be activated into "high" (1) or "low" (0) states by spurious voltage sources if left floating.

For this reason, it is inadvisable to allow a CMOS logic gate input to float under any circumstances. Please note that this is very different from the behaviour of a TTL gate where a floating input was safely interpreted as a "high" (1) logic level.

This may cause a problem if the input to a CMOS logic gate is driven by a single-throw switch, where one state has the input solidly connected to either V_{dd} or ground and the other state has the input floating (not connected to anything):

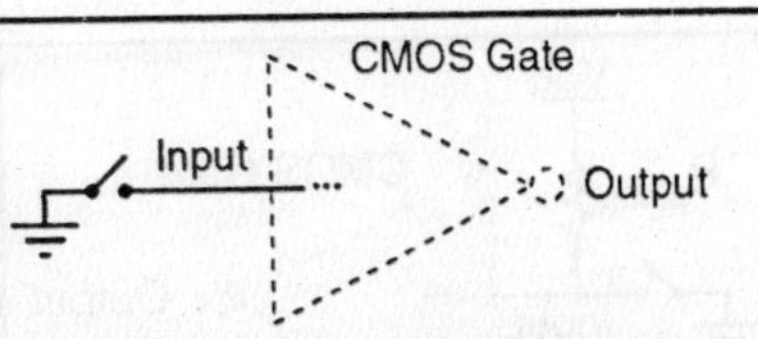

When Switch is closed,the Gate sees a Definite "low"(0)Input. However, when Switch is open, the input logic level will be Uncertain because it's Floatin.

Also, this problem arises if a CMOS gate input is being driven by an *open-collector* TTL gate. Because such a TTL gate's output floats when it goes "high" (1), the CMOS gate input will be left in an uncertain state:

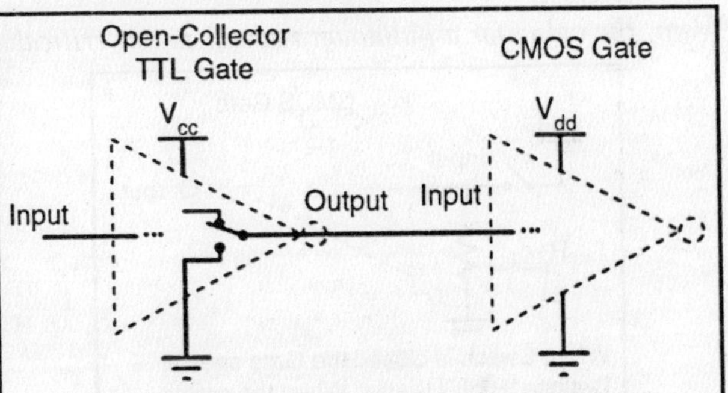

When the Open-Collector TTLGate's Output IS "High" (1),the CMOS Gate's Input will be left Floating and in an Uncertain logic State.

Fortunately, there is an easy solution to this dilemma, one that is used frequently in CMOS logic circuitry. Whenever a single-throw switch (or any other sort of gate output incapable of *both* sourcing and sinking current) is being used to drive a CMOS input, a resistor connected to either V_{dd} or ground may be used to provide a stable logic level for the state in which the driving device's output is floating.

This resistor's value is not critical: 10 kÙ is usually sufficient. When used to provide a "high" (1) logic level in the event of a floating signal source, this resistor is known as a *pullup resistor*:

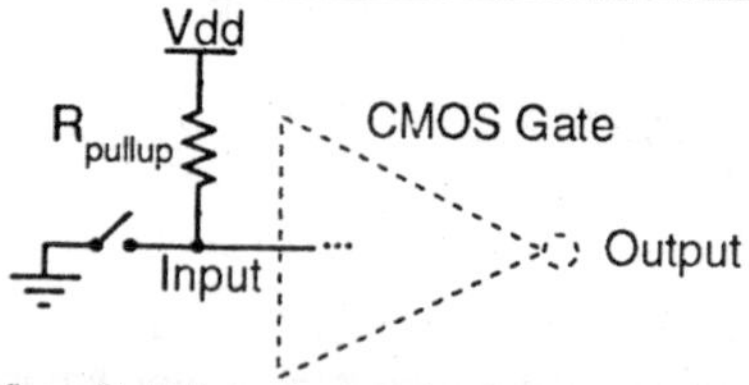

When Switch is closed, the Gate sees a definite "Low" (0)Input. When the Switch is open, R_{pullup} will provide the Connection to Vdd needed to secure a Reliable "High" logic level for the CMOS Gate Input.

When such a resistor is used to provide a "low" (0) logic level in the event of a floating signal source, it is known as a*pulldown resistor.*

Again, the value for a pulldown resistor is not critical:

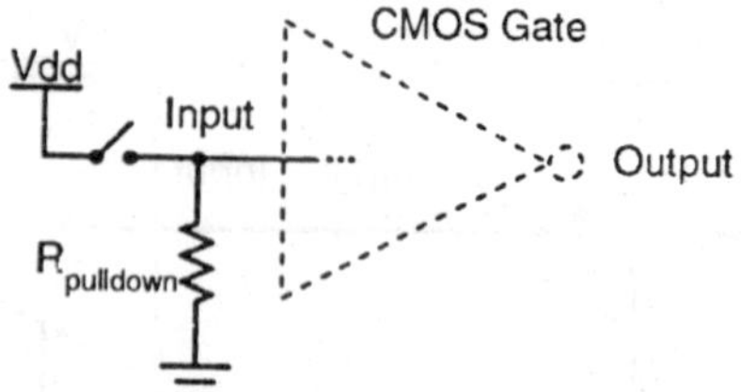

When Switch is closed,the Gate sees a Definite "High"(1)input. When the switch is open,$R_{pulldown}$ will provide the Connection to ground needed to secure a Reliable "Low" logic level for the CMOS Gate input

Because open-collector TTL outputs always sink, never source, current, pull*up* resistors are necessary when interfacing such an output to a CMOS gate input:

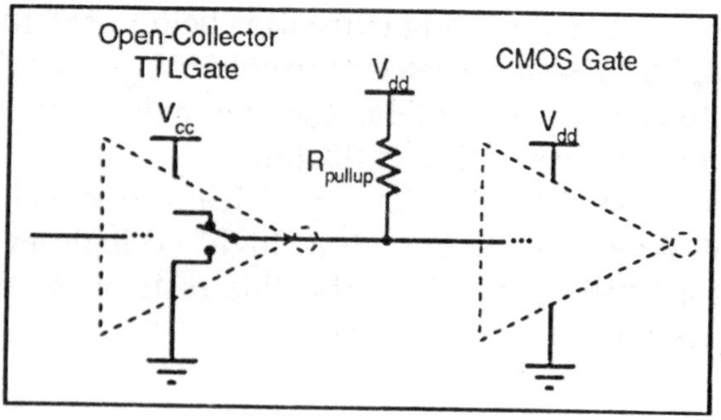

Although the CMOS gates used in the preceding examples were all inverters (single-input), the same principle of pullup and pulldown resistors applies to multiple-input CMOS gates. Of course, a separate pullup or pulldown resistor will be required for each gate input:

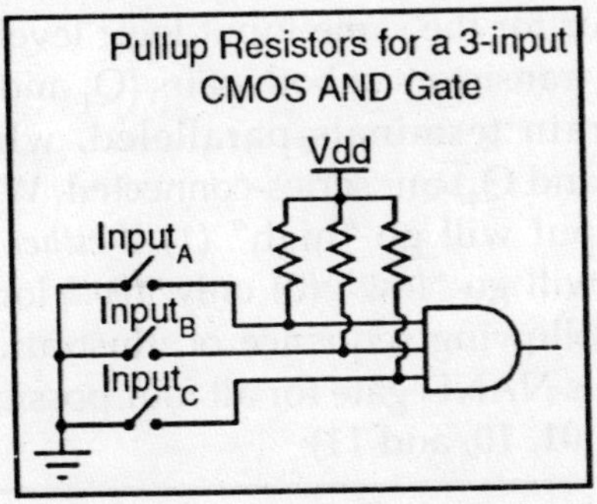

This brings us to the next question: how do we design multiple-input CMOS gates such as AND, NAND, OR, and NOR? Not surprisingly, the answer(s) to this question reveal a simplicity of design much like that of the CMOS inverter over its TTL equivalent.

For example, here is the schematic diagram for a CMOS NAND gate:

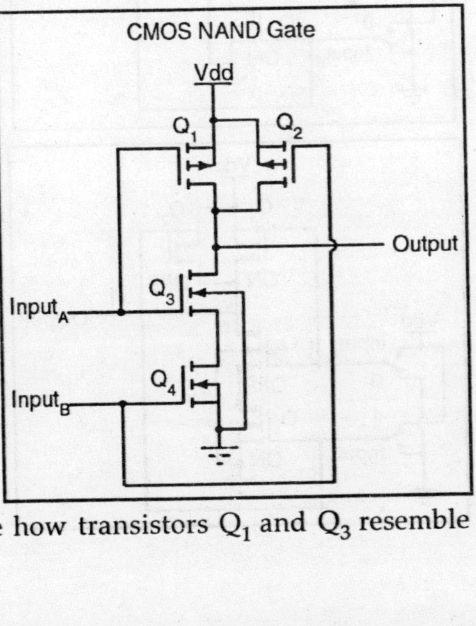

Notice how transistors Q_1 and Q_3 resemble the series-connected complementary pair from the inverter circuit. Both are controlled by the same input signal (input A), the upper transistor turning off and the lower transistor turning on when the input is "high" (1), and visa-versa. Notice also how transistors Q_2 and Q_4 are similarly controlled by the same input signal (input B), and how they will also exhibit the same on/off behaviour for the same input logic levels.

The upper transistors of both pairs (Q_1 and Q_2) have their source and drain terminals paralleled, while the lower transistors (Q_3 and Q_4) are series-connected. What this means is that the output will go "high" (1) if *either* top transistor saturates, and will go "low" (0) only if*both* lower transistors saturate. The following sequence of illustrations shows the behaviour of this NAND gate for all four possibilities of input logic levels (00, 01, 10, and 11):

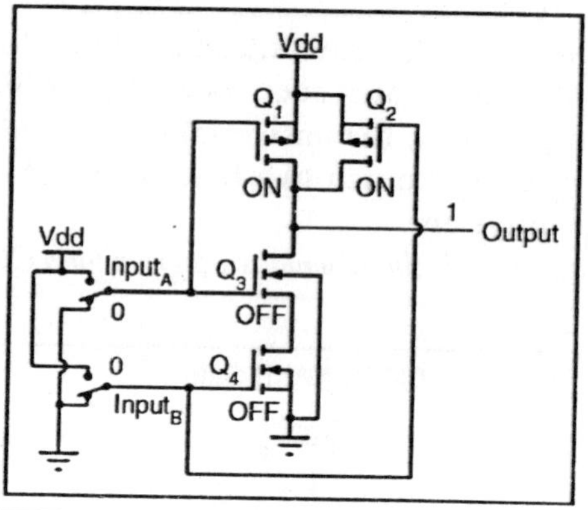

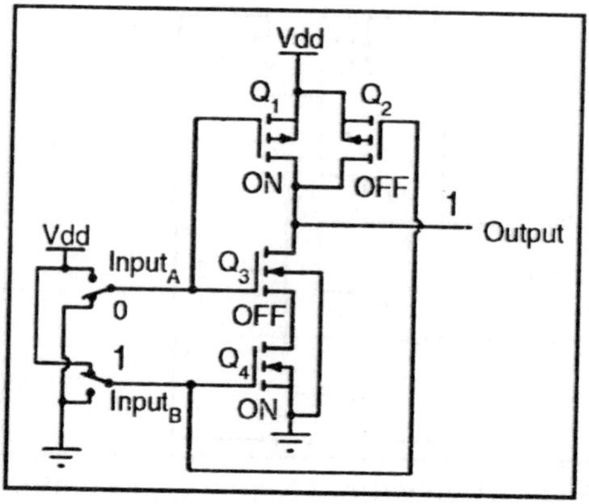

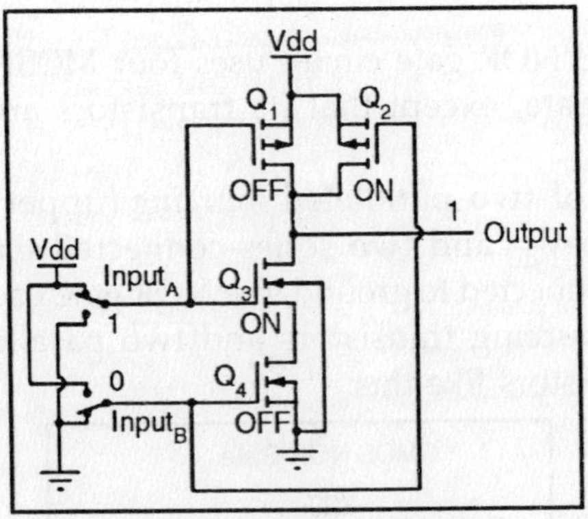

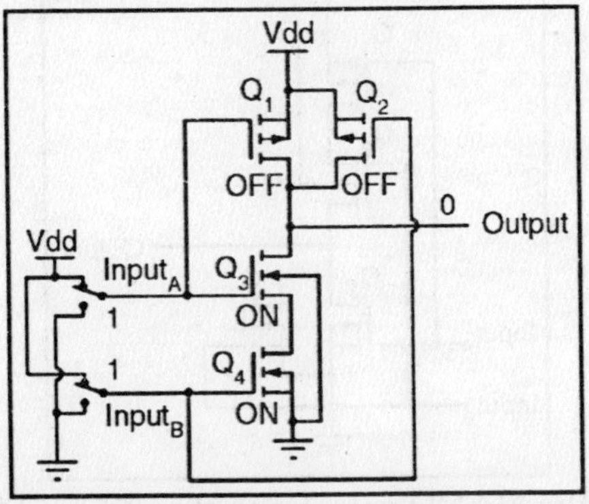

As with the TTL NAND gate, the CMOS NAND gate circuit may be used as the starting point for the creation of an AND gate. All that needs to be added is another stage of transistors to invert the output signal:

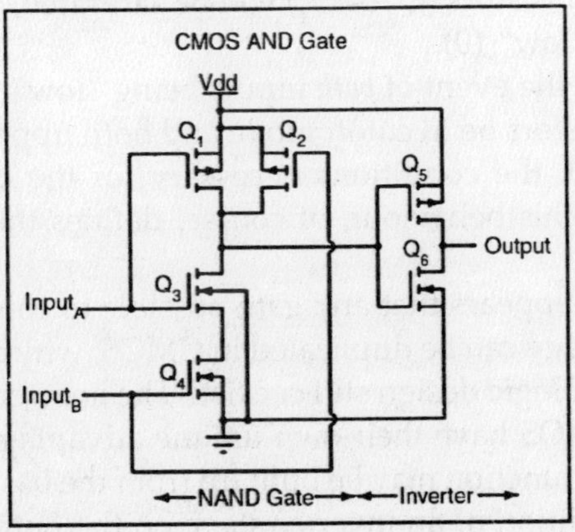

A CMOS NOR gate circuit uses four MOSFETs just like the NAND gate, except that its transistors are differently arranged.

Instead of two paralleled *sourcing* (upper) transistors connected to V_{dd} and two series-connected *sinking* (lower) transistors connected to ground, the NOR gate uses two series-connected sourcing transistors and two parallel-connected sinking transistors like this:

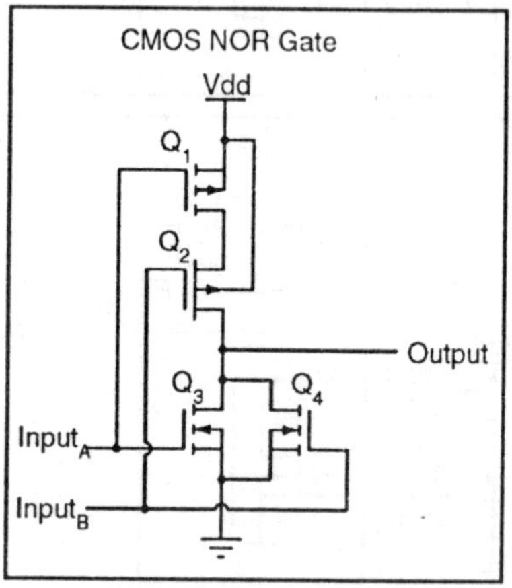

As with the NAND gate, transistors Q_1 and Q_3 work as a complementary pair, as do transistors Q_2 and Q_4. Each pair is controlled by a single input signal.

If *either* input A *or* input B are "high" (1), at least one of the lower transistors (Q_3 or Q_4) will be saturated, thus making the output "low" (0).

Only in the event of *both* inputs being "low" (0) will both lower transistors be in cutoff mode and both upper transistors be saturated, the conditions necessary for the output to go "high" (1). This behaviour, of course, defines the NOR logic function.

Since it appears that any gate possible to construct using TTL technology can be duplicated in CMOS, why do these two "families" of logic design still coexist? The answer is that both TTL and CMOS have their own unique advantages.

The OR function may be built up from the basic NOR gate with the addition of an inverter stage on the output:

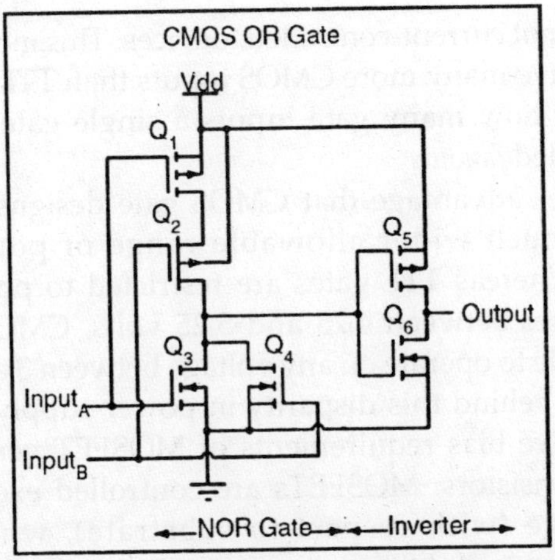

First and foremost on the list of comparisons between TTL and CMOS is the issue of power consumption. In this measure of performance, CMOS is the unchallenged victor. Because the complementary P- and N-channel MOSFET pairs of a CMOS gate circuit are (ideally) never conducting at the same time, there is little or no current drawn by the circuit from the V_{dd} power supply except for what current is necessary to source current to a load. TTL, on the other hand, cannot function without some current drawn at all times, due to the biasing requirements of the bipolar transistors from which it is made.

There is a caveat to this advantage, though. While the power dissipation of a TTL gate remains rather constant regardless of its operating state(s), a CMOS gate dissipates more power as the frequency of its input signal(s) rises. If a CMOS gate is operated in a static (unchanging) condition, it dissipates zero power (ideally). However, CMOS gate circuits draw transient current during every output state switch from "low" to "high" and visa-versa. So, the more often a CMOS gate switches modes, the more often it will draw current from the V_{dd} supply, hence greater power dissipation at greater frequencies.

A CMOS gate also draws much less current from a driving gate output than a TTL gate because MOSFETs are voltage-controlled, not current-controlled, devices. This means that one gate can drive many more CMOS inputs than TTL inputs. The measure of how many gate inputs a single gate output can drive is called *fanout*.

Another advantage that CMOS gate designs enjoy over TTL is a much wider allowable range of power supply voltages. Whereas TTL gates are restricted to power supply (V_{cc}) voltages between 4.75 and 5.25 volts, CMOS gates are typically able to operate on any voltage between 3 and 15 volts! The reason behind this disparity in power supply voltages is the respective bias requirements of MOSFET versus bipolar junction transistors. MOSFETs are controlled exclusively by gate voltage (with respect to substrate), whereas BJTs are *current-controlled* devices.

TTL gate circuit resistances are precisely calculated for proper bias currents assuming a 5 volt regulated power supply. Any significant variations in that power supply voltage will result in the transistor bias currents being incorrect, which then results in unreliable (unpredictable) operation. The only effect that variations in power supply voltage have on a CMOS gate is the voltage definition of a "high" (1) state. For a CMOS gate operating at 15 volts of power supply voltage (V_{dd}), an input signal must be close to 15 volts in order to be considered "high" (1). The voltage threshold for a "low" (0) signal remains the same: near 0 volts.

One decided disadvantage of CMOS is slow speed, as compared to TTL. The input capacitances of a CMOS gate are much, much greater than that of a comparable TTL gate — owing to the use of MOSFETs rather than BJTs — and so a CMOS gate will be slower to respond to a signal transition (low-to-high or visa-versa) than a TTL gate, all other factors being equal.

The RC time constant formed by circuit resistances and the input capacitance of the gate tend to impede the fast rise- and fall-times of a digital logic level, thereby degrading high-frequency performance.

A strategy for minimizing this inherent disadvantage of CMOS gate circuitry is to "buffer" the output signal with additional transistor stages, to increase the overall voltage gain of the device. This provides a faster-transitioning output voltage (high-to-low or low-to-high) for an input voltage slowly changing from one logic state to another. Consider this example, of an "unbuffered" NOR gate versus a "buffered," or *B-series*, NOR gate:

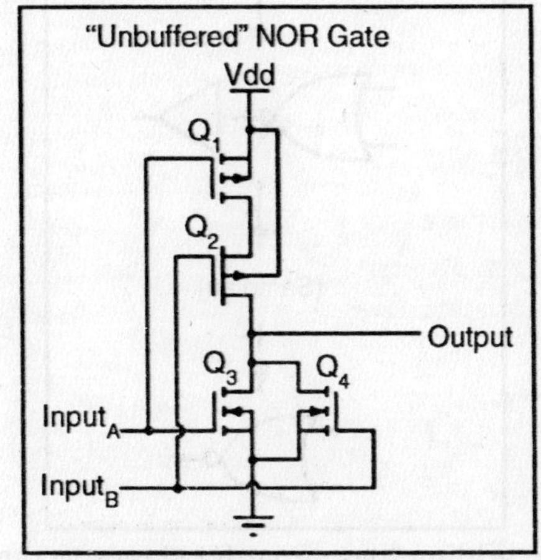

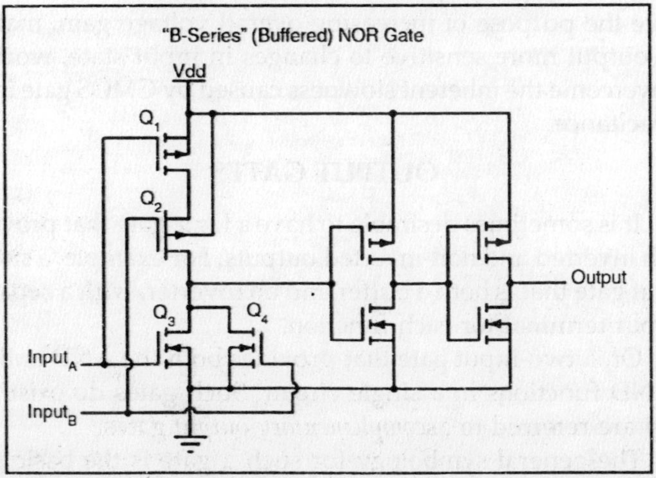

In essence, the B-series design enhancement adds two inverters to the output of a simple NOR circuit. This serves no purpose as far as digital logic is concerned, since two cascaded inverters simply cancel:

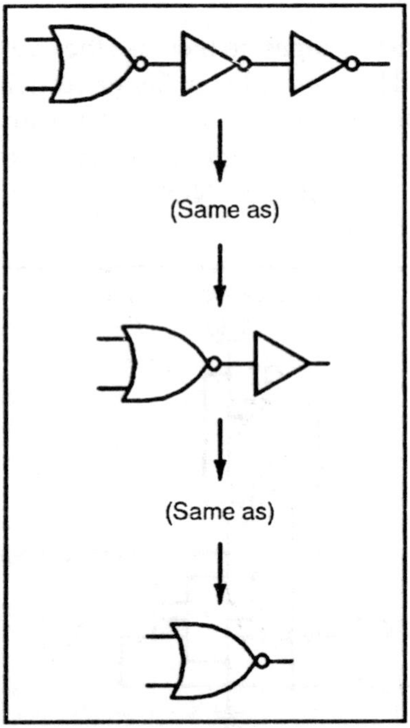

However, adding these inverter stages to the circuit does serve the purpose of increasing overall voltage gain, making the output more sensitive to changes in input state, working to overcome the inherent slowness caused by CMOS gate input capacitance.

OUTPUT GATES

It is sometimes desirable to have a logic gate that provides both inverted and non-inverted outputs. For example, a single-input gate that is both a buffer and an inverter, with a separate output terminal for each function.

Or, a two-input gate that provides both the AND and the NAND functions in a single circuit. Such gates do exist and they are referred to as*complementary output* gates.

The general symbology for such a gate is the basic gate figure with a bar and two output lines protruding from it. An array of complementary gate symbols is shown in the following illustration:

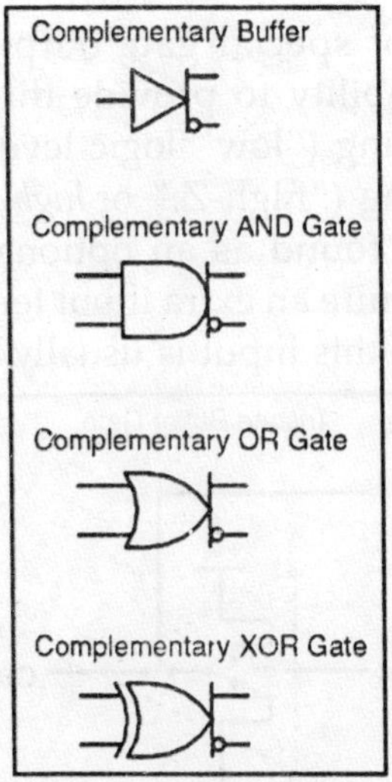

Complementary gates are especially useful in "crowded" circuits where there may not be enough physical room to mount the additional integrated circuit chips necessary to provide both inverted and noninverted outputs using standard gates and additional inverters. They are also useful in applications where a complementary output is necessary from a gate, but the addition of an inverter would introduce an unwanted time lag in the inverted output relative to the noninverted output. The internal circuitry of complemented gates is such that both inverted and noninverted outputs change state at almost exactly the same time:

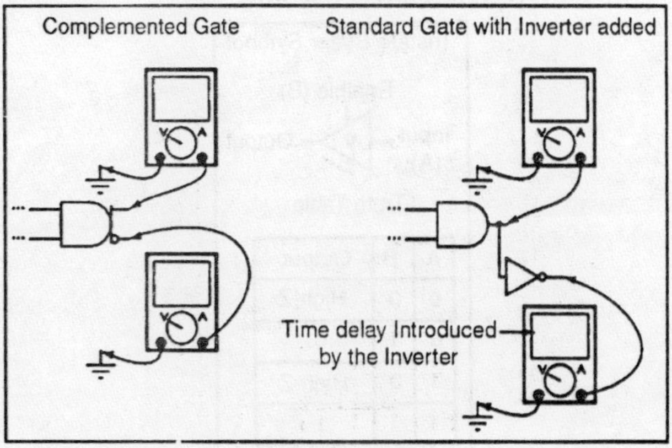

Another type of special gate output is called *tristate*, because it has the ability to provide three different output modes: current sinking ("low" logic level), current sourcing ("high"), and floating ("high-Z," or *high-impedance*). Tristate outputs are usually found as an optional feature on buffer gates. Such gates require an extra input terminal to control the "high-Z" mode, and this input is usually called the *enable*.

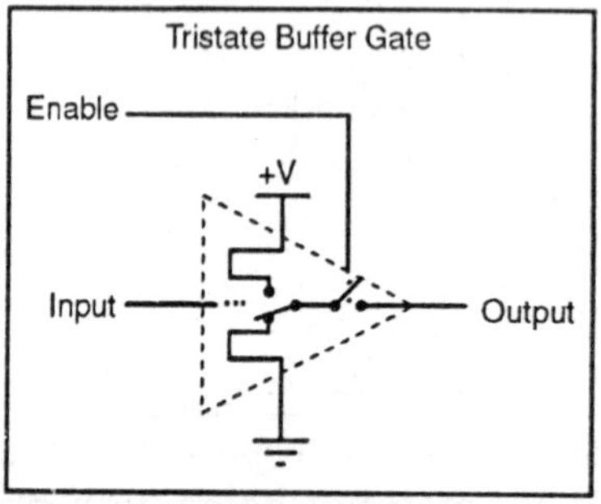

With the enable input held "high" (1), the buffer acts like an ordinary buffer with a totem pole output stage: it is capable of both sourcing and sinking current. However, the output terminal floats (goes into "high-Z" mode) if ever the enable input is grounded ("low"), regardless of the data signal's logic level. In other words, making the enable input terminal "low" (0) effectively *disconnects* the gate from whatever its output is wired to so that it can no longer have any effect.

Tristate buffers are marked in schematic diagrams by a triangle character within the gate symbol like this:

Tristate Buffer Symbol

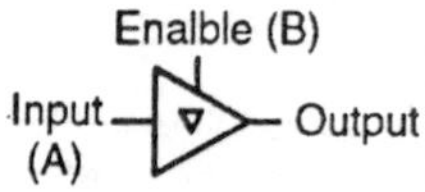

Truth Table

A	B	Output
0	0	High-Z
0	1	0
1	0	High-Z
1	1	1

Tristate buffers are also made with inverted enable inputs. Such a gate acts normal when the enable input is "low" (0) and goes into high-Z output mode when the enable input is "high" (1):

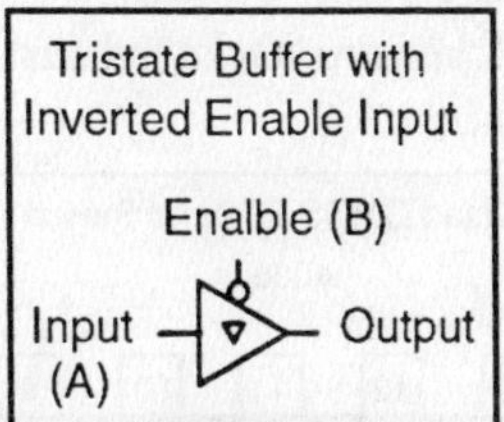

Truth Table

A	B	Output
0	0	0
0	1	High-Z
1	0	1
1	1	High-Z

One special type of gate known as the *bilateral switch* uses gate-controlled MOSFET transistors acting as on/off switches to switch electrical signals, analog or digital. The "on" resistance of such a switch is in the range of several hundred ohms, the "off" resistance being in the range of several hundred *mega*-ohms.

Bilateral switches appear in schematics as SPST (Single-Pole, Single-Throw) switches inside of rectangular boxes, with a control terminal on one of the box's long sides:

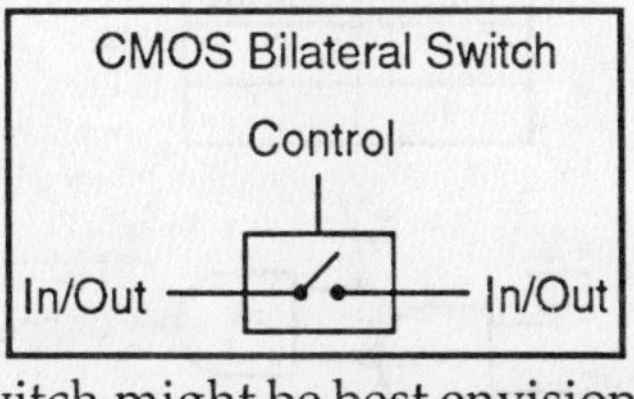

A bilateral switch might be best envisioned as a solid-state (semiconductor) version of an electromechanical relay: a signal-actuated switch contact that may be used to conduct virtually any type of electric signal. Of course, being solid-state, the bilateral switch has none of the undesirable characteristics of electromechanical relays, such as contact "bouncing,"

arcing, slow speed, or susceptibility to mechanical vibration. Conversely, though, they are rather limited in their current-carrying ability. Additionally, the signal conducted by the "contact" must not exceed the power supply "rail" voltages powering the bilateral switch circuit.

Four bilateral switches are packaged inside the popular model "4066" integrated circuit:

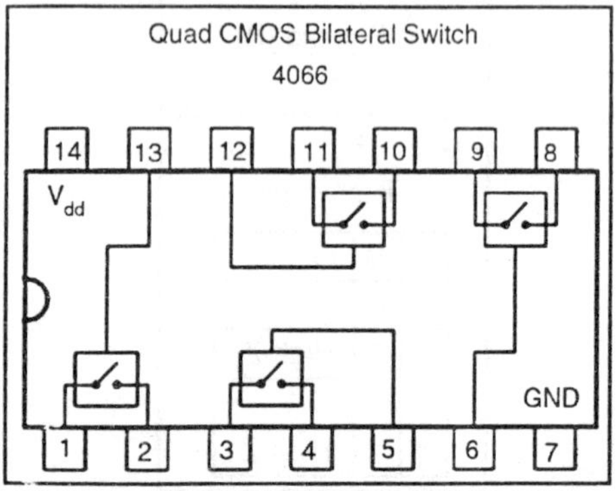

BUFFER FUNCTION

Being that it is quite easy to employ NAND and NOR gates to perform the inverter (NOT) function, it stands to reason that two such stages of gates will result in a buffer function, where the output is the same logical state as the input.

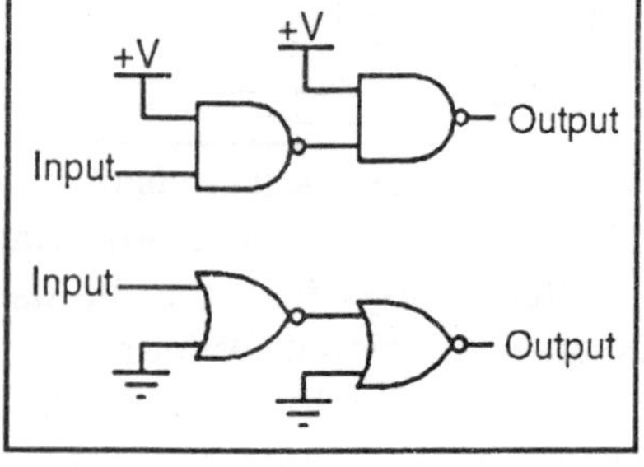

CONSTRUCTING THE AND FUNCTION

To make the AND function from NAND gates, all that is needed is an inverter (NOT) stage on the output of a NAND gate.

This extra inversion "cancels out" the first *N* in *NAND*, leaving the AND function.

It takes a little more work to wrestle the same functionality out of NOR gates, but it can be done by inverting ("NOT") all of the inputs to a NOR gate.

2-Input AND Gate

A	B	Output
0	0	0
0	1	0
1	0	0
1	1	1

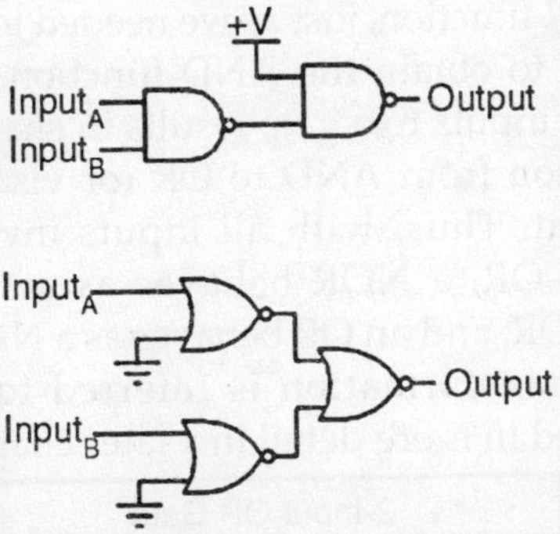

NAND FUNCTION

It would be pointless to show you how to "construct" the NAND function using a NAND gate, since there is nothing to do. To make a NOR gate perform the NAND function, we must invert all inputs to the NOR gate as well as the NOR gate's output. For a two-input gate, this requires three more NOR gates connected as inverters.

2-Input AND Gate

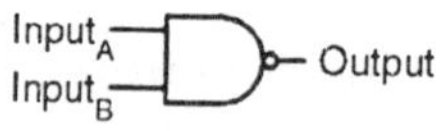

A	B	Output
0	0	1
0	1	1
1	0	1
1	1	0

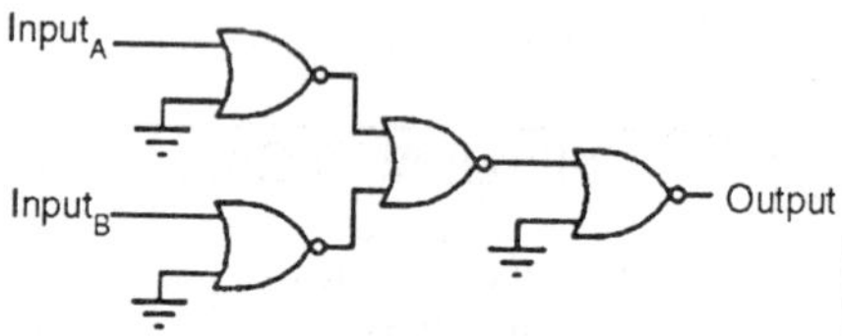

OR FUNCTION

Inverting the output of a NOR gate (with another NOR gate connected as an inverter) results in the OR function. The NAND gate, on the other hand, requires inversion of all inputs to mimic the OR function, just as we needed to invert all inputs of a NOR gate to obtain the AND function. Remember that inversion of all inputs to a gate results in changing that gate's essential function from AND to OR (or visa-versa), plus an inverted output. Thus, with all inputs inverted, a NAND behaves as an OR, a NOR behaves as an AND, an AND behaves as a NOR, and an OR behaves as a NAND. In Boolean algebra, this transformation is referred to as *DeMorgan's Theorem*, covered in more detail in a later chapter of this book.

2-Input OR Gate

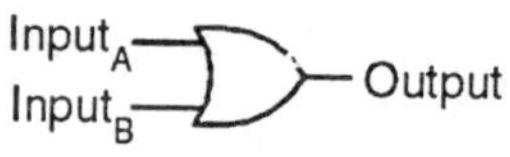

A	B	Output
0	0	0
0	1	1
1	0	1
1	1	1

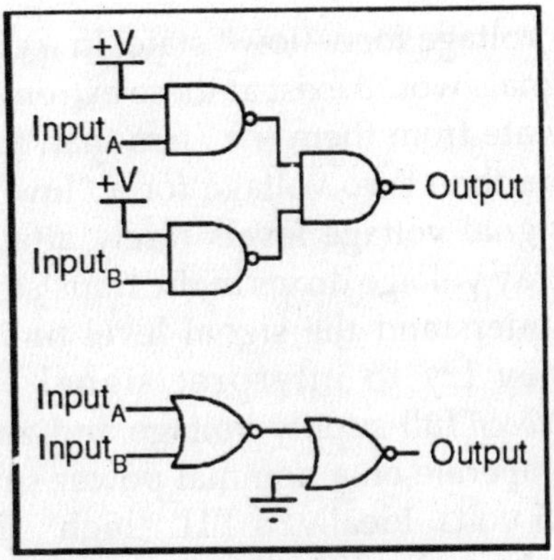

NOR FUNCTION

Much the same as the procedure for making a NOR gate behave as a NAND, we must invert all inputs and the output to make a NAND gate function as a NOR.

2-Input NOR Gate

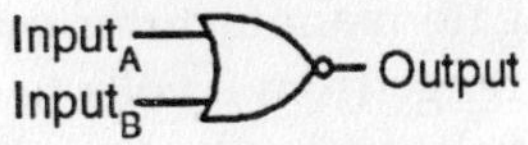

A	B	Output
0	0	1
0	1	0
1	0	0
1	1	0

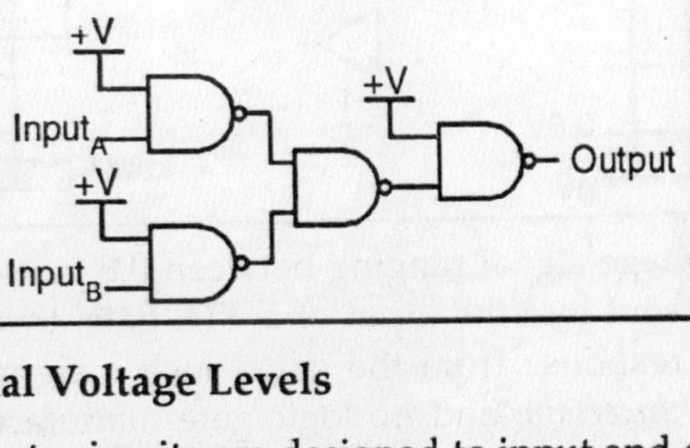

Logic Signal Voltage Levels

Logic gate circuits are designed to input and output only two types of signals: "high" (1) and "low" (0), as represented by a variable voltage: full power supply voltage for a "high" state and zero voltage for a "low" state. In a perfect world, all logic circuit signals would exist at these extreme voltage limits, and never deviate from them (i.e., less than full voltage for a "high," or more than zero voltage for a "low"). However, in reality, logic signal voltage levels rarely attain these perfect limits due to stray voltage drops in the transistor circuitry, and so we must understand the signal level limitations of gate circuits as they try to interpret signal voltages lying somewhere *between* full supply voltage and zero.

TTL gates operate on a nominal power supply voltage of 5 volts, +/- 0.25 volts. Ideally, a TTL "high" signal would be 5.00 volts exactly, and a TTL "low" signal 0.00 volts exactly. However, real TTL gate circuits cannot output such perfect voltage levels, and are designed to accept "high" and "low" signals deviating substantially from these ideal values. "Acceptable" input signal voltages range from 0 volts to 0.8 volts for a "low" logic state, and 2 volts to 5 volts for a "high" logic state. "Acceptable" output signal voltages (voltage levels guaranteed by the gate manufacturer over a specified range of load conditions) range from 0 volts to 0.5 volts for a "low" logic state, and 2.7 volts to 5 volts for a "high" logic state:

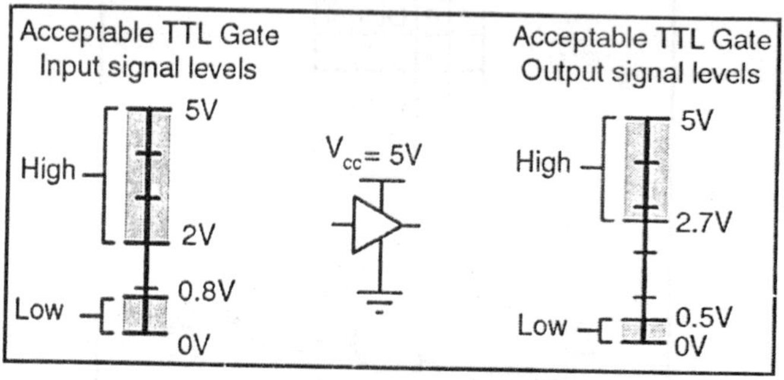

If a voltage signal ranging between 0.8 volts and 2 volts were to be sent into the input of a TTL gate, there would be no certain response from the gate. Such a signal would be considered *uncertain*, and no logic gate manufacturer would guarantee how their gate circuit would interpret such a signal.

As you can see, the tolerable ranges for output signal levels are narrower than for input signal levels, to ensure that any TTL gate outputting a digital signal into the input of another TTL gate will transmit voltages acceptable to the receiving gate. The difference between the tolerable output and input ranges is called the *noise margin* of the gate. For TTL gates, the low-level noise margin is the difference between 0.8 volts and 0.5 volts (0.3 volts), while the high-level noise margin is the difference between 2.7 volts and 2 volts (0.7 volts).

Simply put, the noise margin is the peak amount of spurious or "noise" voltage that may be superimposed on a weak gate output voltage signal before the receiving gate might interpret it wrongly:

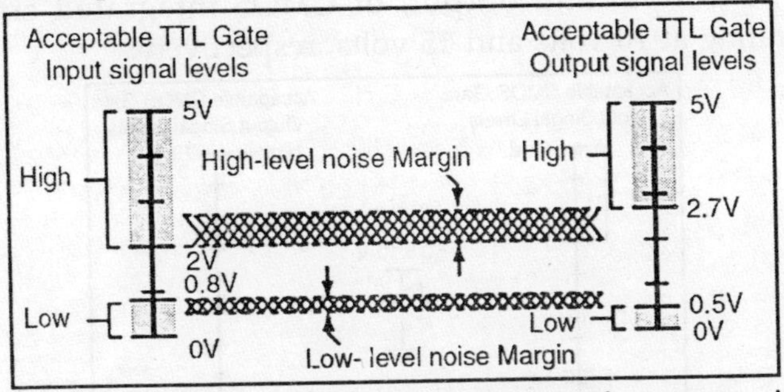

CMOS gate circuits have input and output signal specifications that are quite different from TTL. For a CMOS gate operating at a power supply voltage of 5 volts, the acceptable input signal voltages range from 0 volts to 1.5 volts for a "low" logic state, and 3.5 volts to 5 volts for a "high" logic state. "Acceptable" output signal voltages (voltage levels guaranteed by the gate manufacturer over a specified range of load conditions) range from 0 volts to 0.05 volts for a "low" logic state, and 4.95 volts to 5 volts for a "high" logic state:

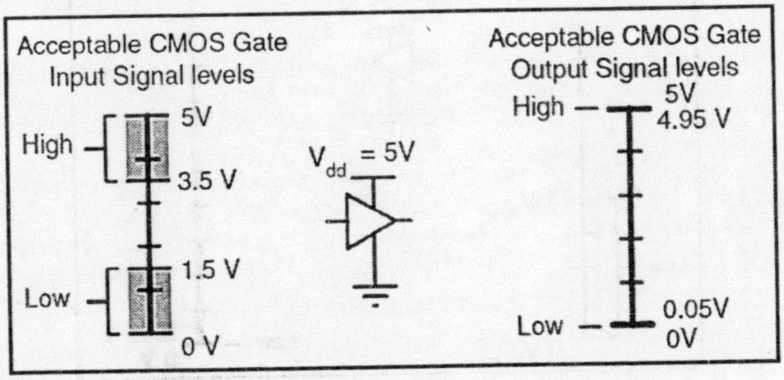

It should be obvious from these figures that CMOS gate circuits have far greater noise margins than TTL: 1.45 volts for CMOS low-level and high-level margins, versus a maximum of 0.7 volts for TTL. In other words, CMOS circuits can tolerate over twice the amount of superimposed "noise" voltage on their input lines before signal interpretation errors will result. CMOS noise margins widen even further with higher operating voltages. Unlike TTL, which is restricted to a power supply voltage of 5 volts, CMOS may be powered by voltages as high as 15 volts (some CMOS circuits as high as 18 volts). Shown here are the acceptable "high" and "low" states, for both input and output, of CMOS integrated circuits operating at 10 volts and 15 volts, respectively:

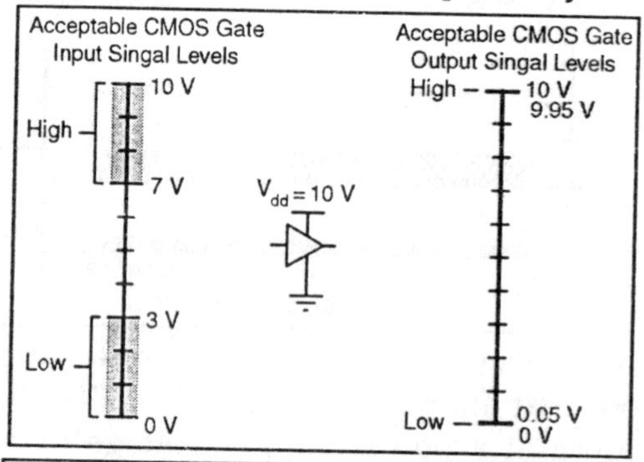

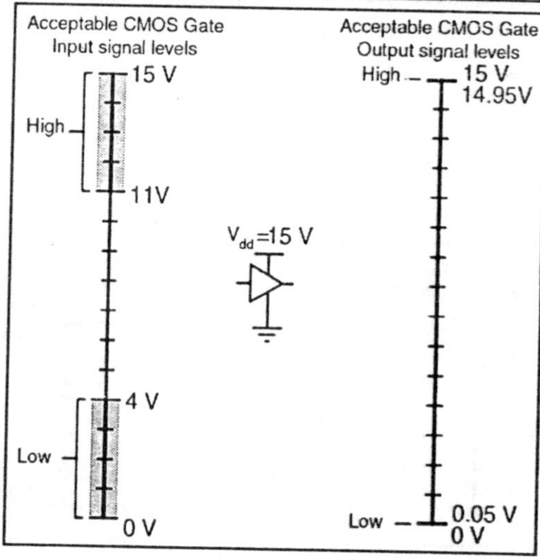

The margins for acceptable "high" and "low" signals may be greater than what is shown in the previous illustrations. What is shown represents "worst-case" input signal performance, based on manufacturer's specifications. In practice, it may be found that a gate circuit will tolerate "high" signals of considerably less voltage and "low" signals of considerably greater voltage than those specified here.

Conversely, the extremely small output margins shown – guaranteeing output states for "high" and "low" signals to within 0.05 volts of the power supply "rails" – are optimistic. Such "solid" output voltage levels will be true only for conditions of minimum loading.

If the gate is sourcing or sinking substantial current to a load, the output voltage will not be able to maintain these optimum levels, due to internal channel resistance of the gate's final output MOSFETs.

Within the "uncertain" range for any gate input, there will be some point of demarcation dividing the gate's actual "low" input signal range from its actual "high" input signal range. That is, somewhere between the lowest "high" signal voltage level and the highest "low" signal voltage level guaranteed by the gate manufacturer, there is a threshold voltage at which the gate will*actually* switch its interpretation of a signal from "low" or "high" or visa-versa. For most gate circuits, this unspecified voltage is a single point:

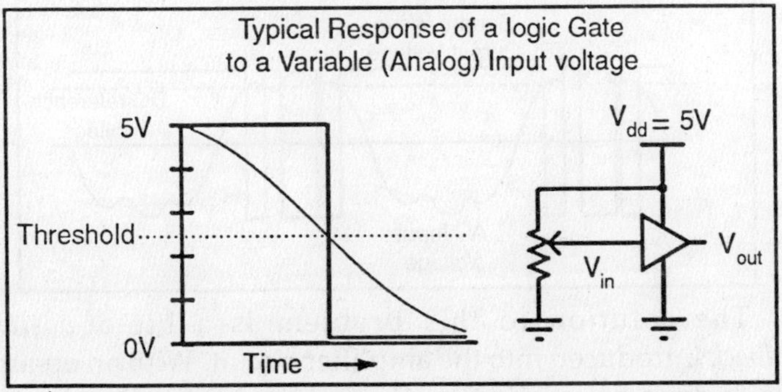

In the presence of AC "noise" voltage superimposed on the DC input signal, a single threshold point at which the gate alters its interpretation of logic level will result in an erratic output:

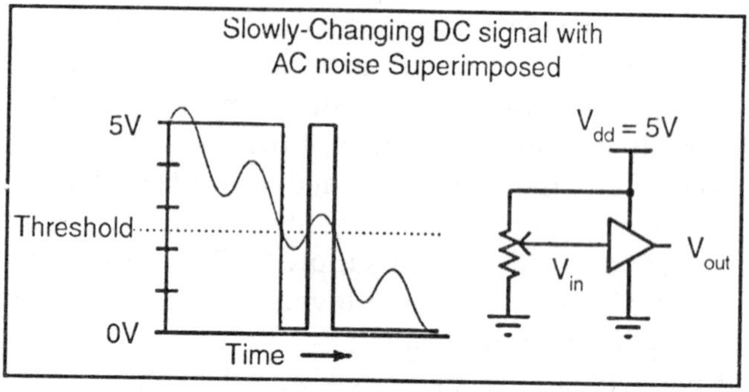

If this scenario looks familiar to you, it's because you remember a similar problem with (analog) voltage comparator op-amp circuits. With a single threshold point at which an input causes the output to switch between "high" and "low" states, the presence of significant noise will cause erratic changes in the output:

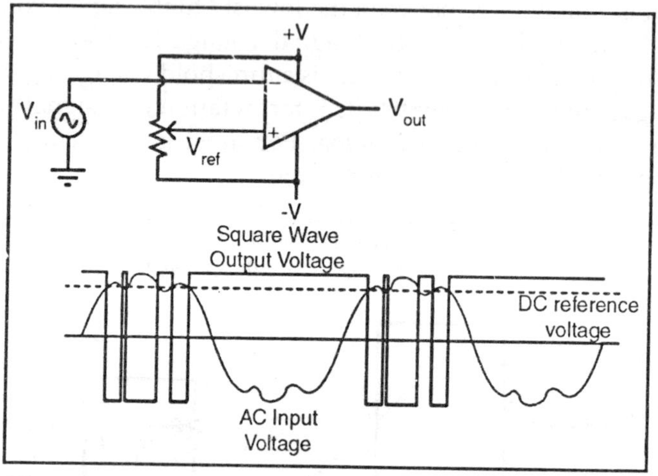

The solution to this problem is a bit of *positive feedback* introduced into the amplifier circuit. With an op-amp, this is done by connecting the output back around to the noninverting (+) input through a resistor. In a gate circuit, this entails redesigning the internal gate circuitry, establishing the feedback inside the gate package rather than through external connections.

A gate so designed is called a *Schmitt trigger*. Schmitt triggers interpret varying input voltages according to *two* threshold voltages: a *positive-going* threshold (V_{T+}), and a *negative-going* threshold (V_{T-}):

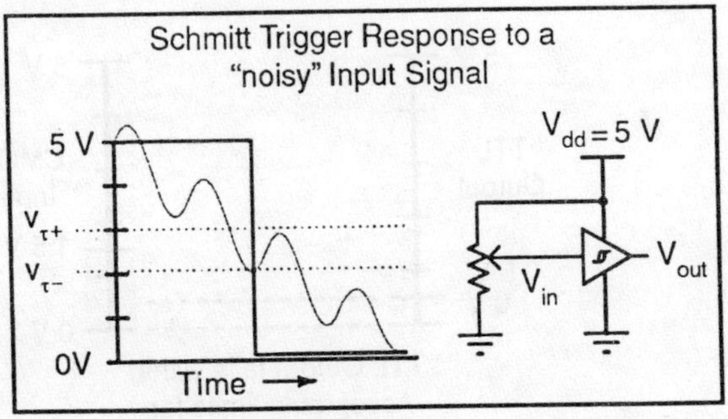

Schmitt trigger gates are distinguished in schematic diagrams by the small "hysteresis" symbol drawn within them, reminiscent of the B-H curve for a ferromagnetic material. Hysteresis engendered by positive feedback within the gate circuitry adds an additional level of noise immunity to the gate's performance.

Schmitt trigger gates are frequently used in applications where noise is expected on the input signal line(s), and/or where an erratic output would be very detrimental to system performance.

The differing voltage level requirements of TTL and CMOS technology present problems when the two types of gates are used in the same system. Although operating CMOS gates on the same 5.00 volt power supply voltage required by the TTL gates is no problem, TTL output voltage levels will not be compatible with CMOS input voltage requirements.

Take for instance a TTL NAND gate outputting a signal into the input of a CMOS inverter gate. Both gates are powered by the same 5.00 volt supply (V_{cc}). If the TTL gate outputs a "low" signal (guaranteed to be between 0 volts and 0.5 volts), it will be properly interpreted by the CMOS gate's input as a "low" (expecting a voltage between 0 volts and 1.5 volts):

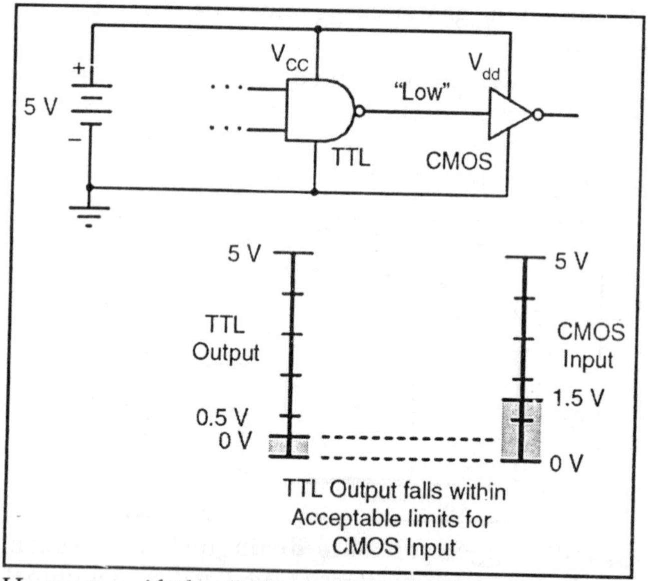

However, if the TTL gate outputs a "high" signal (guaranteed to be between 5 volts and 2.7 volts), it *might not* be properly interpreted by the CMOS gate's input as a "high" (expecting a voltage between 5 volts and *3.5* volts):

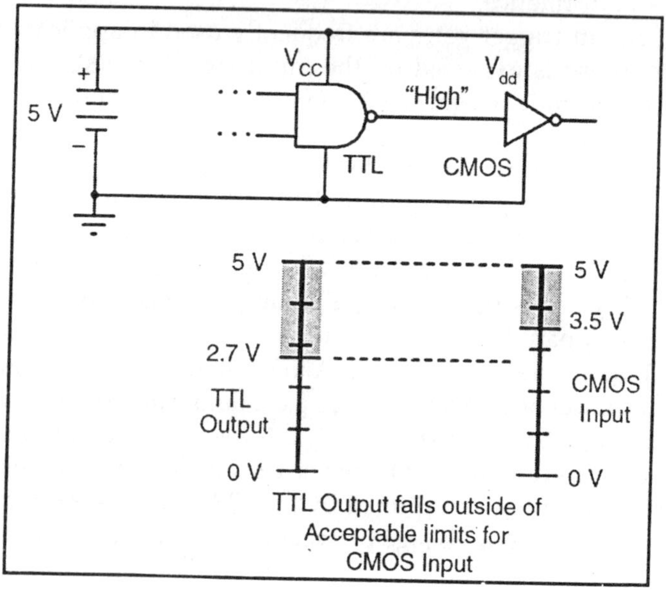

Given this mismatch, it is entirely possible for the TTL gate to output a valid "high" signal (valid, that is, according to the standards for TTL) that lies within the "uncertain" range for the CMOS input, and may be (falsely) interpreted as a "low" by the receiving gate. An easy "fix" for this problem is to augment the TTL gate's "high" signal voltage level by means of a pullup resistor:

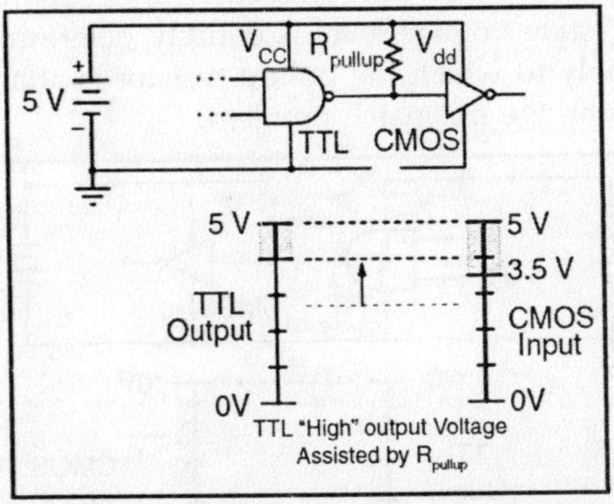

TTL "High" output Voltage Assisted by R_{pullup}

Something more than this, though, is required to interface a TTL output with a CMOS input, if the receiving CMOS gate is powered by a greater power supply voltage:

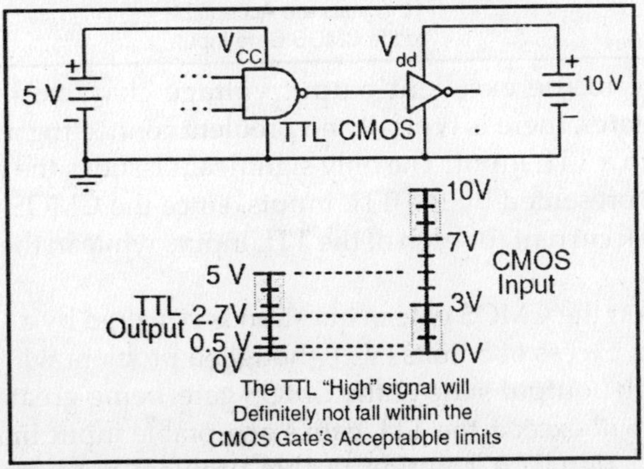

The TTL "High" signal will Definitely not fall within the CMOS Gate's Acceptabble limits

There will be no problem with the CMOS gate interpreting the TTL gate's "low" output, of course, but a "high" signal from the TTL gate is another matter entirely. The guaranteed output voltage range of 2.7 volts to 5 volts from the TTL gate output is nowhere near the CMOS gate's acceptable range of 7 volts to 10 volts for a "high" signal.

If we use an *open-collector* TTL gate instead of a totem-pole output gate, though, a pullup resistor to the 10 volt V_{dd} supply rail will raise the TTL gate's "high" output voltage to the full power supply voltage supplying the CMOS gate. Since an open-collector gate can only sink current, not source current, the "high" state voltage level is entirely determined by the power supply to which the pullup resistor is attached, thus neatly solving the mismatch problem:

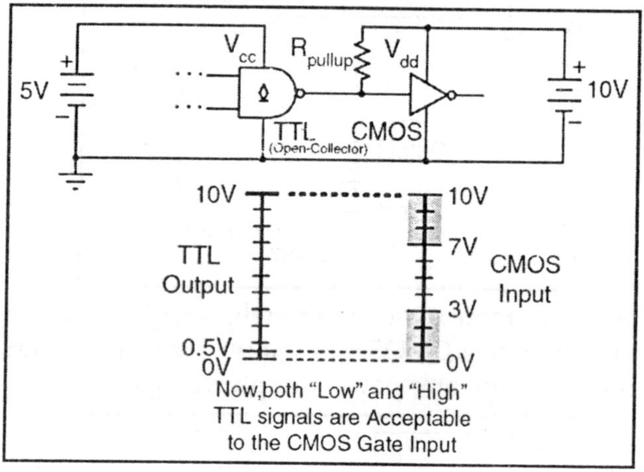

Due to the excellent output voltage characteristics of CMOS gates, there is typically no problem connecting a CMOS output to a TTL input. The only significant issue is the current loading presented by the TTL inputs, since the CMOS output must sink current for each of the TTL inputs while in the "low" state.

When the CMOS gate in question is powered by a voltage source in excess of 5 volts (V_{cc}), though, a problem will result. The "high" output state of the CMOS gate, being greater than 5 volts, will exceed the TTL gate's acceptable input limits for a "high" signal. A solution to this problem is to create an "open-collector" inverter circuit using a discrete NPN transistor, and use it to interface the two gates together:

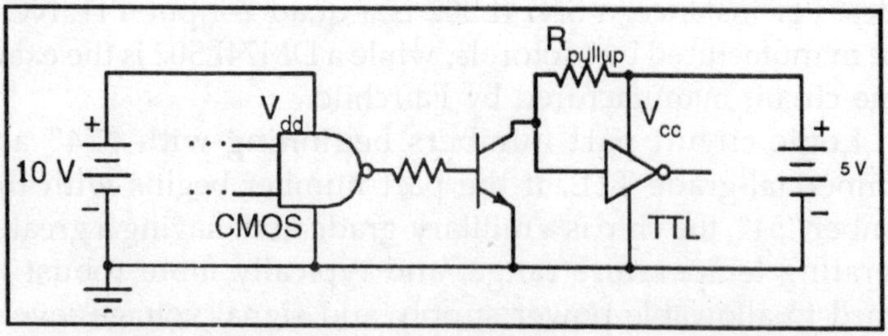

The "R_{pullup}" resistor is optional, since TTL inputs automatically assume a "high" state when left floating, which is what will happen when the CMOS gate output is "low" and the transistor cuts off. Of course, one very important consequence of implementing this solution is the logical inversion created by the transistor: when the CMOS gate outputs a "low" signal, the TTL gate sees a "high" input; and when the CMOS gate outputs a "high" signal, the transistor saturates and the TTL gate sees a "low" input. So long as this inversion is accounted for in the logical scheme of the system, all will be well.

DIP GATE PACKAGING

Digital logic gate circuits are manufactured as integrated circuits: all the constituent transistors and resistors built on a single piece of semiconductor material. The engineer, technician, or hobbyist using small numbers of gates will likely find what he or she needs enclosed in a DIP (Dual Inline Package) housing. DIP-enclosed integrated circuits are available with even numbers of pins, located at 0.100 inch intervals from each other for standard circuit board layout compatibility. Pin counts of 8, 14, 16, 18, and 24 are common for DIP "chips."

Part numbers given to these DIP packages specify what type of gates are enclosed, and how many. These part numbers are industry standards, meaning that a "74LS02" manufactured by Motorola will be identical in function to a "74LS02" manufactured by Fairchild or by any other manufacturer. Letter codes prepended to the part number are unique to the manufacturer, and are not industry-standard codes. For instance, a SN74LS02 is a quad 2-input TTL NOR gate manufactured by Motorola, while a DM74LS02 is the exact same circuit manufactured by Fairchild.

Logic circuit part numbers beginning with "74" are commercial-grade TTL. If the part number begins with the number "54", the chip is a military-grade unit: having a greater operating temperature range, and typically more robust in regard to allowable power supply and signal voltage levels. The letters "LS" immediately following the 74/54 prefix indicate "Low-power Schottky" circuitry, using Schottky-barrier diodes and transistors throughout, to decrease power dissipation. Non-Schottky gate circuits consume more power, but are able to operate at higher frequencies due to their faster switching times.

A few of the more common TTL "DIP" circuit packages are shown here for reference:

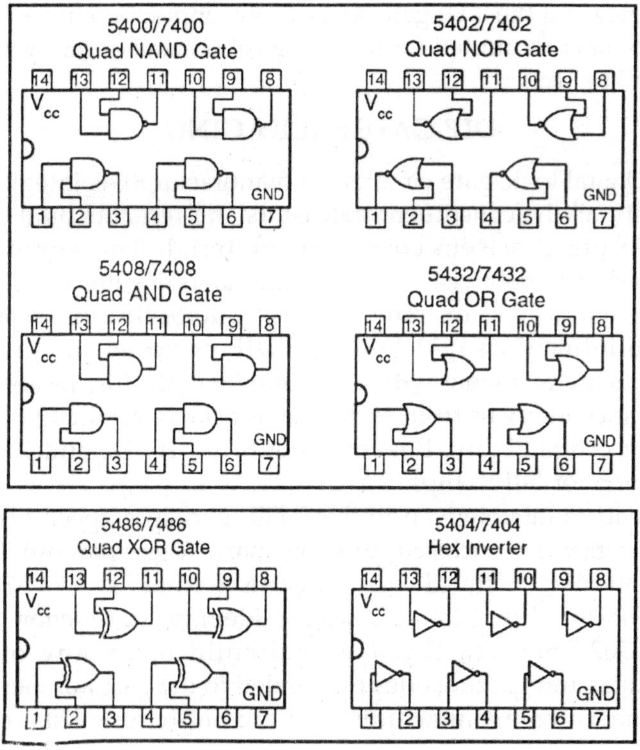

Chapter 4

Switches and Multivibrators

SWITCH

An electrical switch is any device used to interrupt the flow of electrons in a circuit. Switches are essentially binary devices: they are either completely on ("closed") or completely off ("open"). There are many different types of switches, and we will explore some of these types in this chapter.

Though it may seem strange to cover this elementary electrical topic at such a late stage in this book series, I do so because the chapters that follow explore an older realm of digital technology based on mechanical switch contacts rather than solid-state gate circuits, and a thorough understanding of switch types is necessary for the undertaking.

Learning the function of switch-based circuits at the same time that you learn about solid-state logic gates makes both topics easier to grasp, and sets the stage for an enhanced learning experience in Boolean algebra, the mathematics behind digital logic circuits.

The simplest type of switch is one where two electrical conductors are brought in contact with each other by the motion of an actuating mechanism. Other switches are more complex, containing electronic circuits able to turn on or off depending on some physical stimulus (such as light or magnetic field) sensed. In any case, the final output of any switch will be (at least) a pair of wire-connection terminals that will either be connected together by the switch's internal contact mechanism ("closed"), or not connected together ("open").

Any switch designed to be operated by a person is generally called a *hand switch,* and they are manufactured in several varieties:

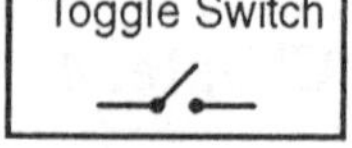

Toggle switches are actuated by a lever angled in one of two or more positions.

The common light switch used in household wiring is an example of a toggle switch. Most toggle switches will come to rest in any of their lever positions, while others have an internal spring mechanism returning the lever to a certain *normal* position, allowing for what is called "momentary" operation.

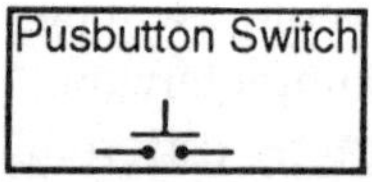

Pushbutton switches are two-position devices actuated with a button that is pressed and released. Most pushbutton switches have an internal spring mechanism returning the button to its "out," or "unpressed," position, for momentary operation. Some pushbutton switches will latch alternately on or off with every push of the button.

Other pushbutton switches will stay in their "in," or "pressed," position until the button is pulled back out. This last type of pushbutton switches usually have a mushroom-shaped button for easy push-pull action.

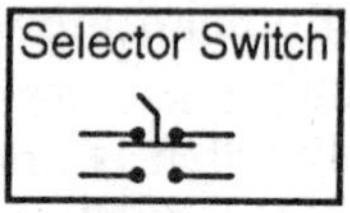

Selector switches are actuated with a rotary knob or lever of some sort to select one of two or more positions. Like the toggle switch, selector switches can either rest in any of their positions or contain spring-return mechanisms for momentary operation.

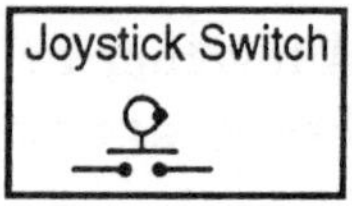

A joystick switch is actuated by a lever free to move in more than one axis of motion. One or more of several switch contact mechanisms are actuated depending on which way the lever is pushed, and sometimes by how *far* it is pushed. The circle-and-dot notation on the switch symbol represents the direction of joystick lever motion required to actuate the contact. Joystick hand switches are commonly used for crane and robot control.

Some switches are specifically designed to be operated by the motion of a machine rather than by the hand of a human operator. These motion-operated switches are commonly called *limit switches,* because they are often used to limit the motion of a machine by turning off the actuating power to a component if it moves too far.

As with hand switches, limit switches come in several varieties:

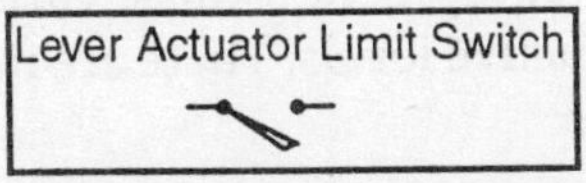

These limit switches closely resemble rugged toggle or selector hand switches fitted with a lever pushed by the machine part. Often, the levers are tipped with a small roller bearing, preventing the lever from being worn off by repeated contact with the machine part.

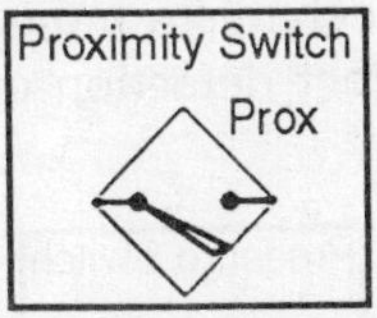

. Proximity switches sense the approach of a metallic machine part either by a magnetic or high-frequency electromagnetic field. Simple proximity switches use a permanent magnet to actuate a sealed switch mechanism whenever the machine part gets close (typically 1 inch or less). More complex proximity switches work like a metal detector, energizing a coil of wire with a high-frequency current, and electronically monitoring the magnitude of that current.

If a metallic part (not necessarily magnetic) gets close enough to the coil, the current will increase, and trip the monitoring circuit. The symbol shown here for the proximity switch is of the electronic variety, as indicated by the diamond-shaped box surrounding the switch. A non-electronic proximity switch would use the same symbol as the lever-actuated limit switch.

Another form of proximity switch is the optical switch, comprised of a light source and photocell. Machine position is detected by either the interruption or reflection of a light beam. Optical switches are also useful in safety applications, where beams of light can be used to detect personnel entry into a dangerous area.

In many industrial processes, it is necessary to monitor various physical quantities with switches. Such switches can be used to sound alarms, indicating that a process variable has exceeded normal parameters, or they can be used to shut down processes or equipment if those variables have reached dangerous or destructive levels. There are many different types of process switches:

These switches sense the rotary speed of a shaft either by a centrifugal weight mechanism mounted on the shaft, or by some kind of non-contact detection of shaft motion such as optical or magnetic.

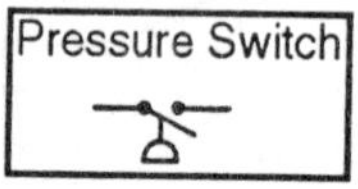

Gas or liquid pressure can be used to actuate a switch mechanism if that pressure is applied to a piston, diaphragm, or bellows, which converts pressure to mechanical force.

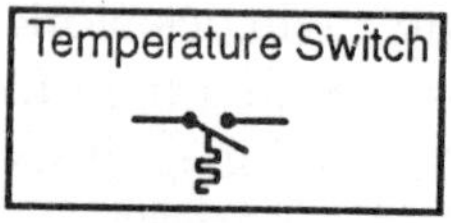

An inexpensive temperature-sensing mechanism is the

"bimetallic strip:" a thin strip of two metals, joined back-to-back, each metal having a different rate of thermal expansion. When the strip heats or cools, differing rates of thermal expansion between the two metals causes it to bend. The bending of the strip can then be used to actuate a switch contact mechanism.

Other temperature switches use a brass bulb filled with either a liquid or gas, with a tiny tube connecting the bulb to a pressure-sensing switch. As the bulb is heated, the gas or liquid expands, generating a pressure increase which then actuates the switch mechanism.

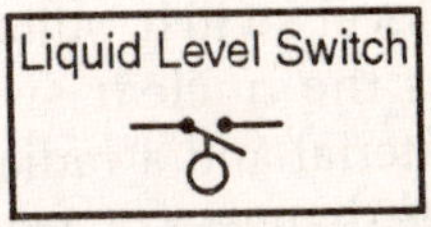

A floating object can be used to actuate a switch mechanism when the liquid level in an tank rises past a certain point. If the liquid is electrically conductive, the liquid itself can be used as a conductor to bridge between two metal probes inserted into the tank at the required depth. The conductivity technique is usually implemented with a special design of relay triggered by a small amount of current through the conductive liquid. In most cases it is impractical and dangerous to switch the full load current of the circuit through a liquid.

Level switches can also be designed to detect the level of solid materials such as wood chips, grain, coal, or animal feed in a storage silo, bin, or hopper. A common design for this application is a small paddle wheel, inserted into the bin at the desired height, which is slowly turned by a small electric motor.

When the solid material fills the bin to that height, the material prevents the paddle wheel from turning. The torque response of the small motor than trips the switch mechanism. Another design uses a "tuning fork" shaped metal prong, inserted into the bin from the outside at the desired height. The fork is vibrated at its resonant frequency by an electronic circuit and magnet/electromagnet coil assembly. When the bin fills to that height, the solid material dampens the vibration of the fork, the change in vibration amplitude and/or frequency detected by the electronic circuit.

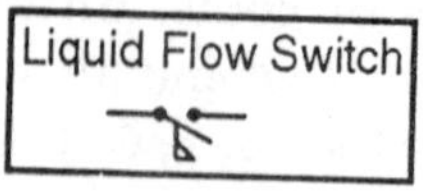

Inserted into a pipe, a flow switch will detect any gas or liquid flow rate in excess of a certain threshold, usually with a small paddle or vane which is pushed by the flow. Other flow switches are constructed as differential pressure switches, measuring the pressure drop across a restriction built into the pipe.

Another type of level switch, suitable for liquid or solid material detection, is the nuclear switch. Composed of a radioactive source material and a radiation detector, the two are mounted across the diameter of a storage vessel for either solid or liquid material. Any height of material beyond the level of the source/detector arrangement will attenuate the strength of radiation reaching the detector. This decrease in radiation at the detector can be used to trigger a relay mechanism to provide a switch contact for measurement, alarm point, or even control of the vessel level.

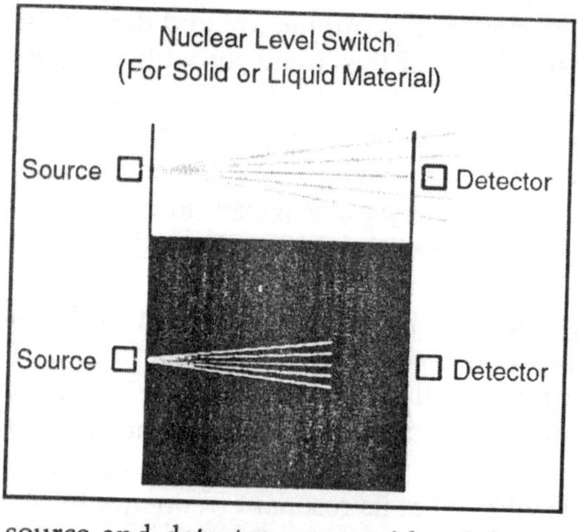

Both source and detector are outside of the vessel, with no intrusion at all except the radiation flux itself. The radioactive sources used are fairly weak and pose no immediate health threat to operations or maintenance personnel.

As usual, there is usually more than one way to implement a switch to monitor a physical process or serve as an operator control. There is usually no single "perfect" switch for any application, although some obviously exhibit certain advantages over others. Switches must be intelligently matched to the task for efficient and reliable operation.

SWITCH CONTACT DESIGN

A switch can be constructed with any mechanism bringing two conductors into contact with each other in a controlled manner. This can be as simple as allowing two copper wires to touch each other by the motion of a lever, or by directly pushing two metal strips into contact. However, a good switch design must be rugged and reliable, and avoid presenting the operator with the possibility of electric shock. Therefore, industrial switch designs are rarely this crude.

The conductive parts in a switch used to make and break the electrical connection are called *contacts*. Contacts are typically made of silver or silver-cadmium alloy, whose conductive properties are not significantly compromised by surface corrosion or oxidation. Gold contacts exhibit the best corrosion resistance, but are limited in current-carrying capacity and may "cold weld" if brought together with high mechanical force. Whatever the choice of metal, the switch contacts are guided by a mechanism ensuring square and even contact, for maximum reliability and minimum resistance.

Contacts such as these can be constructed to handle extremely large amounts of electric current, up to thousands of amps in some cases.

The limiting factors for switch contact ampacity are as follows:

- Heat generated by current through metal contacts (while closed).
- Sparking caused when contacts are opened or closed.
- The voltage across open switch contacts (potential of current jumping across the gap).

One major disadvantage of standard switch contacts is the exposure of the contacts to the surrounding atmosphere. In a nice, clean, control-room environment, this is generally not a problem. However, most industrial environments are not this benign. The presence of corrosive chemicals in the air can cause contacts to deteriorate and fail prematurely. Even more troublesome is the possibility of regular contact sparking causing flammable or explosive chemicals to ignite.

When such environmental concerns exist, other types of contacts can be considered for small switches. These other types of contacts are sealed from contact with the outside air, and therefore do not suffer the same exposure problems that standard contacts do.

A common type of sealed-contact switch is the mercury switch. Mercury is a metallic element, liquid at room temperature. Being a metal, it possesses excellent conductive properties. Being a liquid, it can be brought into contact with metal probes (to close a circuit) inside of a sealed chamber simply by tilting the chamber so that the probes are on the bottom.

Many industrial switches use small glass tubes containing mercury which are tilted one way to close the contact, and tilted another way to open. Aside from the problems of tube breakage and spilling mercury (which is a toxic material), and susceptibility to vibration, these devices are an excellent alternative to open-air switch contacts wherever environmental exposure problems are a concern.

CONTACT "BOUNCE"

When a switch is actuated and contacts touch one another under the force of actuation, they are supposed to establish continuity in a single, crisp moment. Unfortunately, though, switches do not exactly achieve this goal. Due to the mass of the moving contact and any elasticity inherent in the mechanism and/or contact materials, contacts will "bounce" upon closure for a period of milliseconds before coming to a full rest and providing unbroken contact.

In many applications, switch bounce is of no consequence:

it matters little if a switch controlling an incandescent lamp "bounces" for a few cycles every time it is actuated. Since the lamp's warm-up time greatly exceeds the bounce period, no irregularity in lamp operation will result.

However, if the switch is used to send a signal to an electronic amplifier or some other circuit with a fast response time, contact bounce may produce very noticeable and undesired effects:

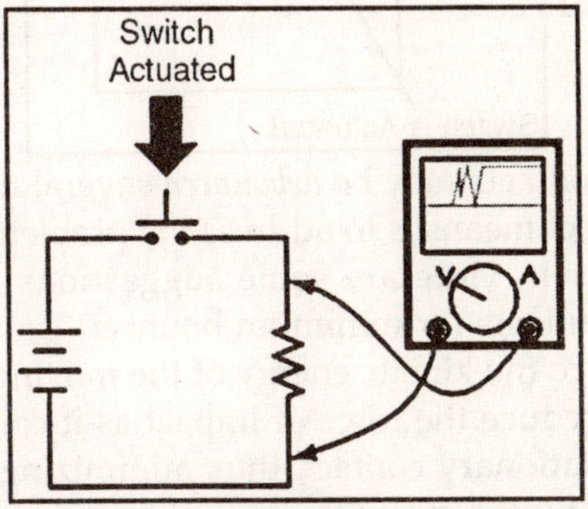

A closer look at the oscilloscope display reveals a rather ugly set of makes and breaks when the switch is actuated a single time:

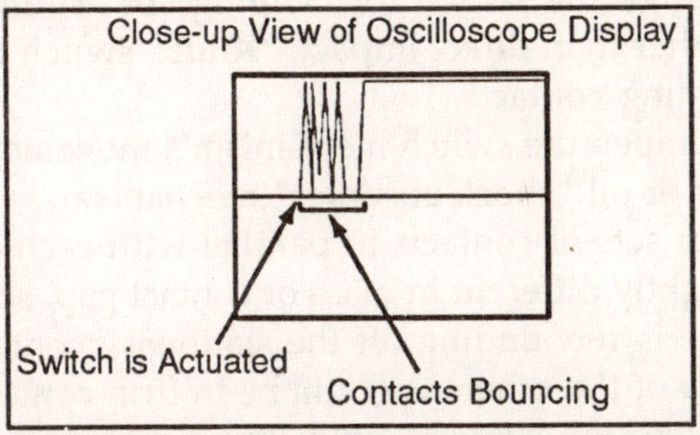

If, for example, this switch is used to provide a "clock" signal to a digital counter circuit, so that each actuation of the pushbutton switch is supposed to increment the counter by a value of 1, what will happen instead is the counter will increment by several counts each time the switch is actuated. Since mechanical switches often interface with digital electronic circuits in modern systems, switch contact bounce is a frequent design consideration. Somehow, the "chattering" produced by bouncing contacts must be eliminated so that the receiving circuit sees a clean, crisp off/on transition:

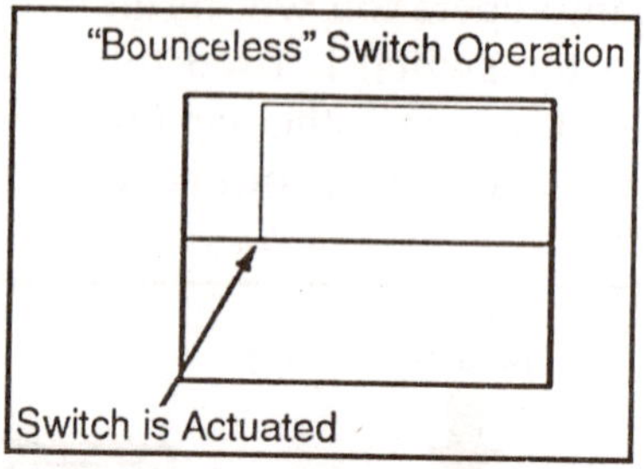

Switch contacts may be *debounced* several different ways. The most direct means is to address the problem at its source: the switch itself. Here are some suggestions for designing switch mechanisms for minimum bounce:

- Reduce the kinetic energy of the moving contact. This will reduce the force of impact as it comes to rest on the stationary contact, thus minimizing bounce.
- Use "buffer springs" on the stationary contact(s) so that they are free to recoil and gently absorb the force of impact from the moving contact.
- Design the switch for "wiping" or "sliding" contact rather than direct impact. "Knife" switch designs use sliding contacts.
- Dampen the switch mechanism's movement using an air or oil "shock absorber" mechanism.
- Use sets of contacts in parallel with each other, each slightly different in mass or contact gap, so that when one is rebounding off the stationary contact, at least one of the others will still be in firm contact.
- "Wet" the contacts with liquid mercury in a sealed environment. After initial contact is made, the surface tension of the mercury will maintain circuit continuity even though the moving contact may bounce off the stationary contact several times.

Each one of these suggestions sacrifices some aspect of switch performance for limited bounce, and so it is impractical to design *all* switches with limited contact bounce in mind.

Alterations made to reduce the kinetic energy of the contact may result in a small open-contact gap or a slow-moving contact, which limits the amount of voltage the switch may handle and the amount of current it may interrupt.

Sliding contacts, while non-bouncing, still produce "noise" (irregular current caused by irregular contact resistance when moving), and suffer from more mechanical wear than normal contacts. Multiple, parallel contacts give less bounce, but only at greater switch complexity and cost. Using mercury to "wet" the contacts is a very effective means of bounce mitigation, but it is unfortunately limited to switch contacts of low ampacity. Also, mercury-wetted contacts are usually limited in mounting position, as gravity may cause the contacts to "bridge" accidently if oriented the wrong way.

If re-designing the switch mechanism is not an option, mechanical switch contacts may be debounced externally, using other circuit components to condition the signal. A low-pass filter circuit attached to the output of the switch, for example, will reduce the voltage/current fluctuations generated by contact bounce:

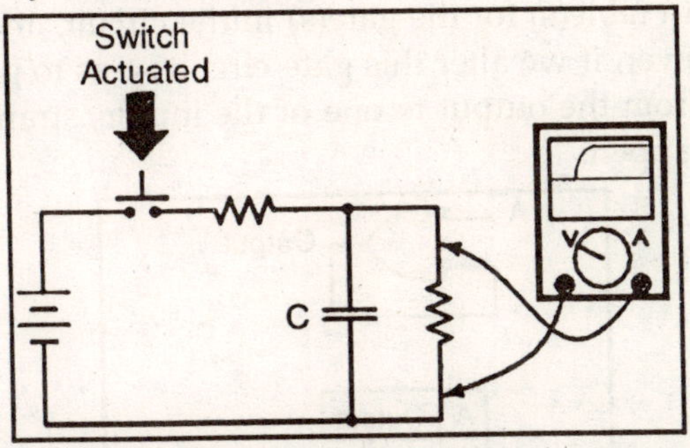

Switch contacts may be debounced electronically, using hysteretic transistor circuits (circuits that "latch" in either a high or a low state) with built-in time delays (called "one-shot" circuits), or two inputs controlled by a double-throw switch.

MULTIVIBRATORS

DIGITAL LOGIC WITH FEEDBACK

With simple gate and combinational logic circuits, there

is a definite output state for any given input state. Take the truth table of an OR gate, for instance:

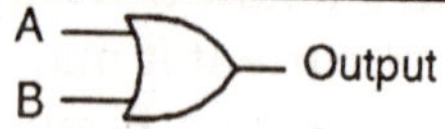

A	B	Output
0	0	0
0	1	1
1	0	1
1	1	1

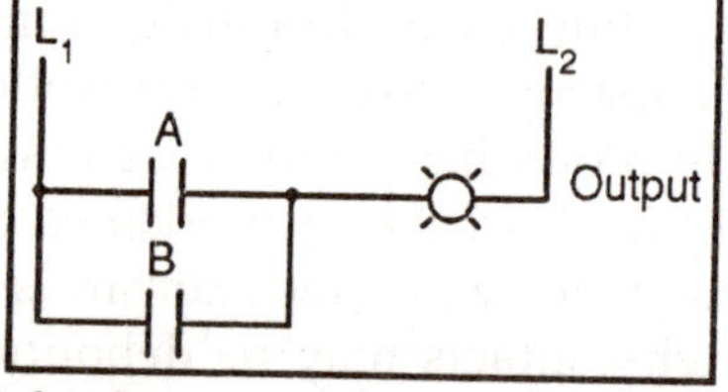

For each of the four possible combinations of input states (0-0, 0-1, 1-0, and 1-1), there is one, definite, unambiguous output state. Whether we're dealing with a multitude of cascaded gates or a single gate, that output state is determined by the truth table(s) for the gate(s) in the circuit, and nothing else. However, if we alter this gate circuit so as to give signal feedback from the output to one of the inputs, strange things begin to happen:

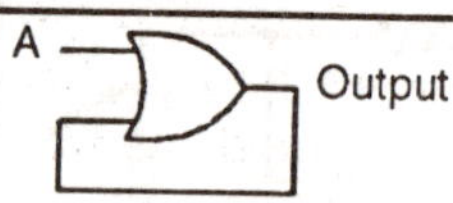

A	Output
0	?
1	1

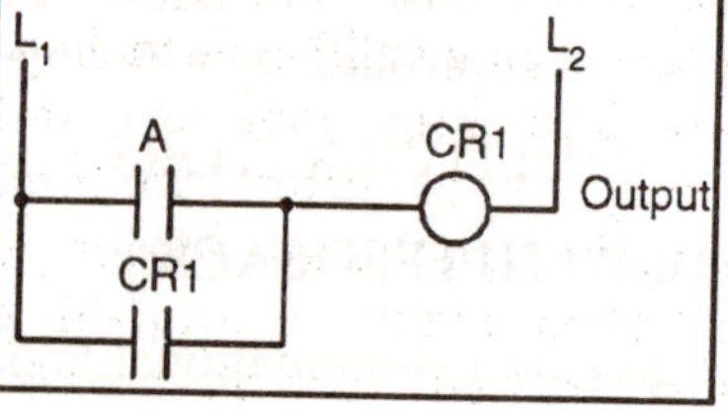

We know that if A is 1, the output *must* be 1, as well. Such is the nature of an OR gate: any "high" (1) input forces the output "high" (1). If A is "low" (0), however, we cannot guarantee the logic level or state of the output in our truth table.

Since the output feeds back to one of the OR gate's inputs, and we know that any 1 input to an OR gates makes the output 1, this circuit will "latch" in the 1 output state after any time that A is 1.

When A is 0, the output could be either 0 or 1,*depending on the circuit's prior state!* The proper way to complete the above truth table would be to insert the word *latch* in place of the question mark, showing that the output maintains its last state when A is 0. Any digital circuit employing feedback is called a *multivibrator*.

The example we just explored with the OR gate was a very simple example of what is called a *bistable* multivibrator. It is called "bistable" because it can hold stable in one of *two*possible output states, either 0 or 1.

There are also *monostable* multivibrators, which have only *one* stable output state (that other state being momentary), which we'll explore later; and *astable* multivibrators, which have no stable state (oscillating back and forth between an output of 0 and 1).

A very simple astable multivibrator is an inverter with the output fed directly back to the input:

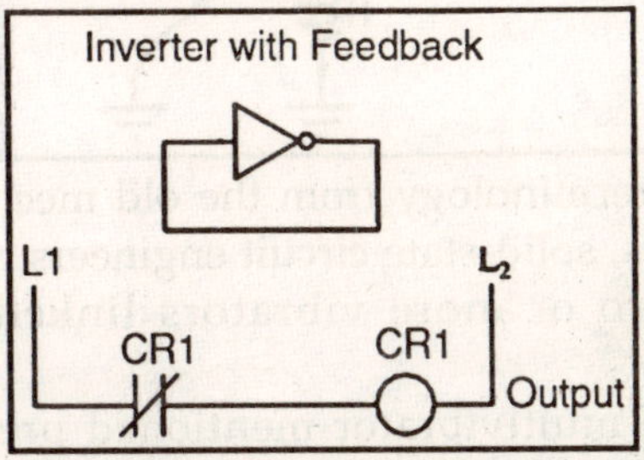

When the input is 0, the output switches to 1. That 1 output gets fed back to the input as a 1. When the input is 1, the output switches to 0. That 0 output gets fed back to the input as a 0, and the cycle repeats itself. The result is a high frequency (several megahertz) oscillator, if implemented with a solid-state (semiconductor) inverter gate:

If implemented with relay logic, the resulting oscillator will be considerably slower, cycling at a frequency well within the audio range.

The *buzzer* or *vibrator* circuit thus formed was used extensively in early radio circuitry, as a way to convert steady, low-voltage DC power into pulsating DC power which could then be stepped up in voltage through a transformer to produce the high voltage necessary for operating the vacuum tube amplifiers.

Henry Ford's engineers also employed the buzzer/transformer circuit to create continuous high voltage for operating the spark plugs on Model T automobile engines:

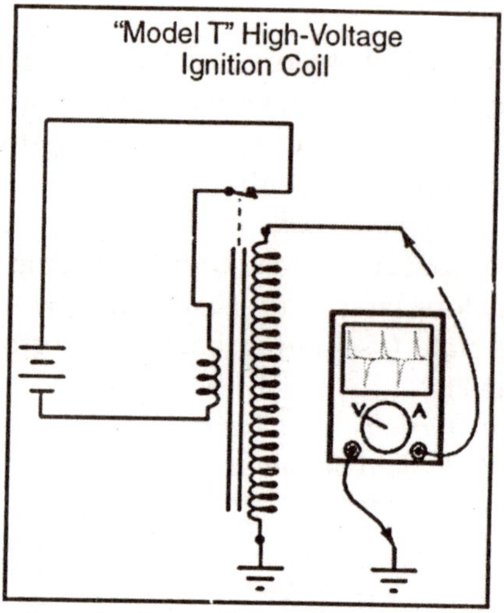

Borrowing terminology from the old mechanical buzzer (vibrator) circuits, solid-state circuit engineers referred to any circuit with two or more vibrators linked together as a *multivibrator*.

The astable multivibrator mentioned previously, with only one "vibrator," is more commonly implemented with multiple gates, as we'll see later.

The most interesting and widely used multivibrators are of the bistable variety, so we'll explore them in detail now.

STABLE STAGES

The S-R Latch

A bistable multivibrator has *two* stable states, as indicated by the prefix *bi* in its name. Typically, one state is referred to as*set* and the other as *reset*. The simplest bistable device, therefore, is known as a *set-reset*, or S-R, latch.

To create an S-R latch, we can wire two NOR gates in such a way that the output of one feeds back to the input of another, and visa-versa, like this:

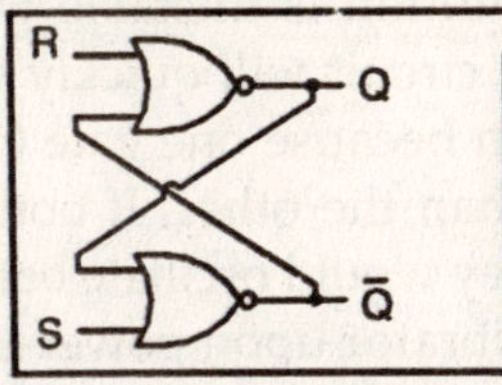

S	R	Q	Q̄
0	0	Latch	Latch
0	1	0	1
1	0	1	0
1	1	0	0

The Q and not-Q outputs are supposed to be in opposite states. I say "supposed to" because making both the S and R inputs equal to 1 results in both Q and not-Q being 0. For this reason, having both S and R equal to 1 is called an *invalid* or*illegal* state for the S-R multivibrator. Otherwise, making S=1 and R=0 "sets" the multivibrator so that Q=1 and not-Q=0. Conversely, making R=1 and S=0 "resets" the multivibrator in the opposite state. When S and R are both equal to 0, the multivibrator's outputs "latch" in their prior states.

Note how the same multivibrator function can be implemented in ladder logic, with the same results:

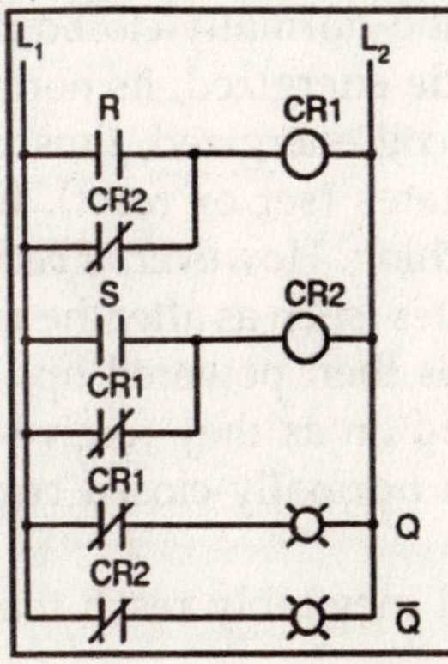

S	R	Q	Q̄
0	0	Latch	Latch
0	1	0	1
1	0	1	0
1	1	0	0

By definition, a condition of Q=1 and not-Q=0 is *set*. A condition of Q=0 and not-Q=1 is *reset*. These terms are universal in describing the output states of any multivibrator circuit.

The astute observer will note that the initial power-up condition of either the gate or ladder variety of S-R latch is such that both gates (coils) start in the de-energized mode. As such, one would expect that the circuit will start up in an invalid condition, with both Q and not-Q outputs being in the same state. Actually, this is true.

However, the invalid condition is unstable with both S and R inputs inactive, and the circuit will quickly stabilize in either the set or reset condition because one gate (or relay) is bound to react a little faster than the other. If both gates (or coils) were *precisely identical*, they would oscillate between high and low like an astable multivibrator upon power-up without ever reaching a point of stability! Fortunately for cases like this, such a precise match of components is a rare possibility.

It must be noted that although an astable (continually oscillating) condition would be extremely rare, there will most likely be a cycle or two of oscillation in the above circuit, and the final state of the circuit (set or reset) after power-up would be unpredictable. The root of the problem is a *race condition* between the two relays CR_1 and CR_2.

A race condition occurs when two mutually-exclusive events are simultaneously initiated through different circuit elements by a single cause. In this case, the circuit elements are relays CR_1 and CR_2, and their de-energized states are mutually exclusive due to the normally-closed interlocking contacts. If one relay coil is de-energized, its normally-closed contact will keep the other coil energized, thus maintaining the circuit in one of two states (set or reset). Interlocking prevents *both* relays from latching. However, if *both* relay coils start in their de-energized states (such as after the whole circuit has been de-energized and is then powered up) both relays will "race" to become latched on as they receive power (the "single cause") through the normally-closed contact of the other relay.

One of those relays will inevitably reach that condition

before the other, thus opening its normally-closed interlocking contact and de-energizing the other relay coil. Which relay "wins" this race is dependent on the physical characteristics of the relays and not the circuit design, so the designer cannot ensure which state the circuit will fall into after power-up.

Race conditions should be avoided in circuit design primarily for the unpredictability that will be created. One way to avoid such a condition is to insert a time-delay relay into the circuit to disable one of the competing relays for a short time, giving the other one a clear advantage. In other words, by purposely slowing down the de-energization of one relay, we ensure that the other relay will always "win" and the race results will always be predictable. Here is an example of how a time-delay relay might be applied to the above circuit to avoid the race condition:

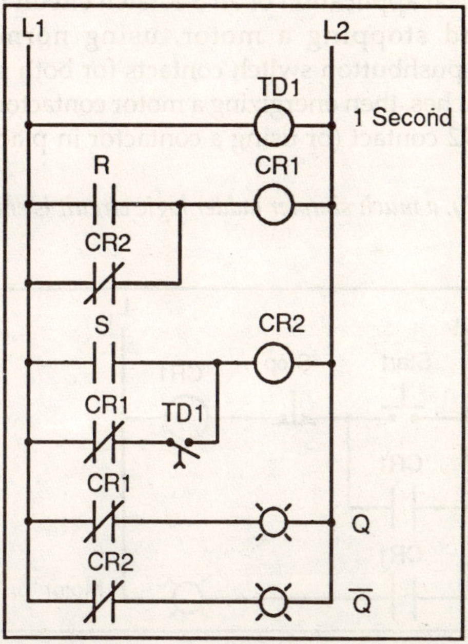

When the circuit powers up, time-delay relay contact TD1 in the fifth rung down will delay closing for 1 second. Having that contact open for 1 second prevents relay CR2 from energizing through contact CR1 in its normally-closed state after power-up. Therefore, relay CR1 will be allowed to energize first (with a 1-second head start), thus opening the normally-closed CR1contact in the fifth rung, preventing CR2 from being energized without the S input going active. The end result is that the circuit powers up cleanly and predictably in the reset state with S=0 and R=0.

It should be mentioned that race conditions are not restricted to relay circuits. Solid-state logic gate circuits may also suffer from the ill effects of race conditions if improperly designed. Complex computer programmes, for that matter, may also incur race problems if improperly designed. Race problems are a possibility for any sequential system, and may not be discovered until some time after initial testing of the system. They can be very difficult problems to detect and eliminate.

A practical application of an S-R latch circuit might be for starting and stopping a motor, using normally-open, momentary pushbutton switch contacts for both start (S) and stop (R) switches, then energizing a motor contactor with either a CR1 or CR2 contact (or using a contactor in place of CR1 or CR2).

Normally, a much simpler ladder logic circuit is employed, such as this:

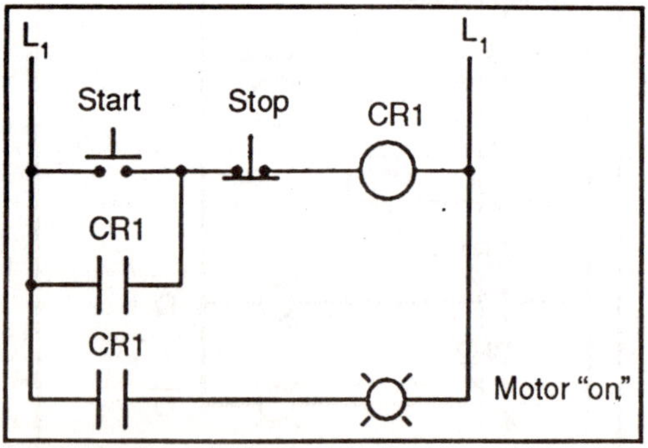

In the above motor start/stop circuit, the CR1 contact in parallel with the start switch contact is referred to as a"seal-in" contact, becausə it"seals" or latches control relay CR1 in the energized state after the start switch has been released. To break the"seal," or to"unlatch" or"reset" the circuit, the stop pushbutton is pressed, which de-energizes CR1 and restores the seal-in contact to its normally open status. Notice, however, that this circuit performs much the same function as the S-R latch. Also note that this circuit has no inherent instability problem (if even a remote possibility) as does the double-relay S-R latch design.

In semiconductor form, S-R latches come in prepackaged units so that you don't have to build them from individual gates.

They are symbolized as such:

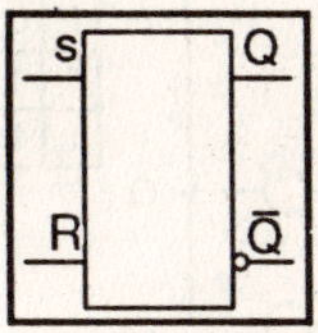

The Gated S-R Latch

It is sometimes useful in logic circuits to have a multivibrator which changes state only when certain conditions are met, regardless of its S and R input states. The conditional input is called the enable, and is symbolized by the letter E.

Study the following example to see how this works:

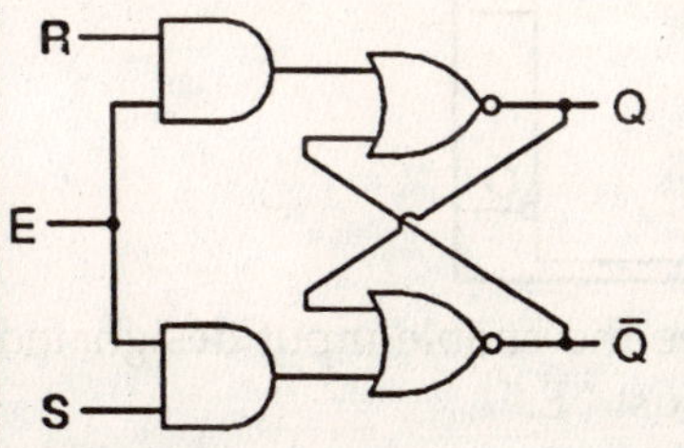

E	S	R	Q	$\overline{Q}$
0	0	0	Latch	Latch
0	0	1	Latch	Latch
0	1	0	Latch	Latch
0	1	1	Latch	Latch
1	0	0	Latch	Latch
1	0	1	0	1
1	1	0	1	0
1	1	1	0	0

When the E=0, the outputs of the two AND gates are forced to 0, regardless of the states of either S or R. Consequently, the circuit behaves as though S and R were both 0, latching the Q and not-Q outputs in their last states. Only when the enable input is activated (1) will the latch respond to the S and R inputs.

Note the identical function in ladder logic:

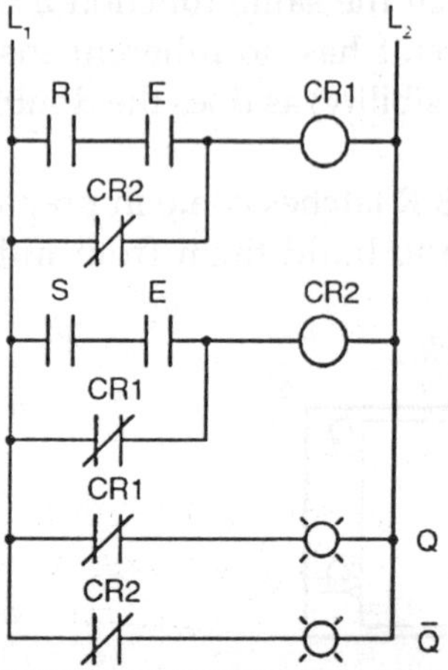

E	S	R	Q	$\bar{Q}$
0	0	0	Latch	Latch
0	0	1	Latch	Latch
0	1	0	Latch	Latch
0	1	1	Latch	Latch
1	0	0	Latch	Latch
1	0	1	0	1
1	1	0	1	0
1	1	1	0	0

A practical application of this might be the same motor control circuit (with two normally-open pushbutton switches for *start*and *stop*), except with the addition of a master lockout input (E) that disables both pushbuttons from having control over the motor when it's low (0). Once again, these multivibrator circuits are available as prepackaged semiconductor devices, and are symbolized as such:

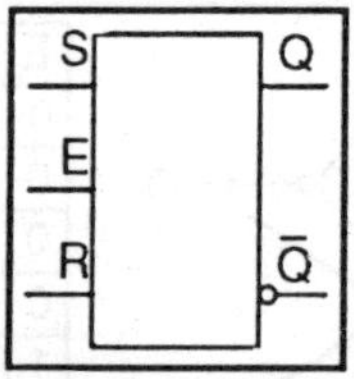

It is also common to see the enable input designated by the letters "EN" instead of just "E."

THE D LATCH

Since the enable input on a gated S-R latch provides a way to latch the Q and not-Q outputs without regard to the status of S or R, we can eliminate one of those inputs to create a multivibrator latch circuit with no "illegal" input states. Such a circuit is called a D latch, and its internal logic looks like this:

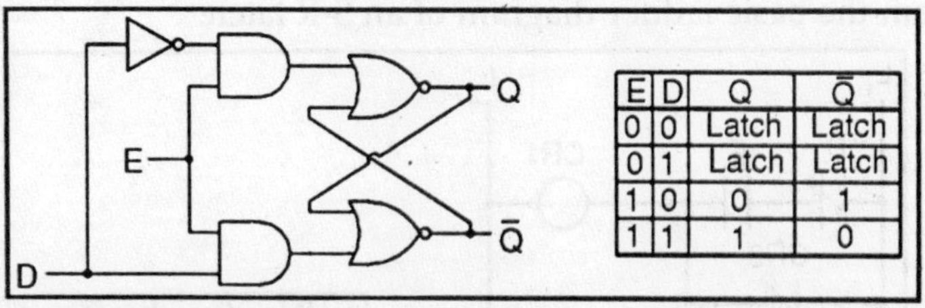

E	D	Q	$\overline{Q}$
0	0	Latch	Latch
0	1	Latch	Latch
1	0	0	1
1	1	1	0

Note that the R input has been replaced with the complement (inversion) of the old S input, and the S input has been renamed to D. As with the gated S-R latch, the D latch will not respond to a signal input if the enable input is 0 — it simply stays latched in its last state. When the enable input is 1, however, the Q output follows the D input.

Since the R input of the S-R circuitry has been done away with, this latch has no "invalid" or "illegal" state. Q and not-Q are*always* opposite of one another. If the above diagram is confusing at all, the next diagram should make the concept simpler:

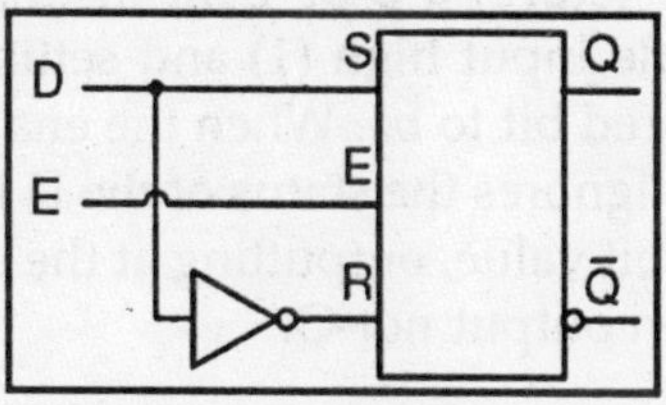

Like both the S-R and gated S-R latches, the D latch circuit may be found as its own prepackaged circuit, complete with a standard symbol:

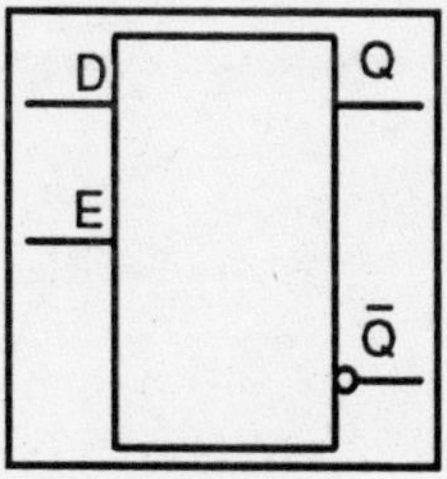

The D latch is nothing more than a gated S-R latch with an inverter added to make R the complement (inverse) of S. Let's explore the ladder logic equivalent of a D latch, modified from the basic ladder diagram of an S-R latch:

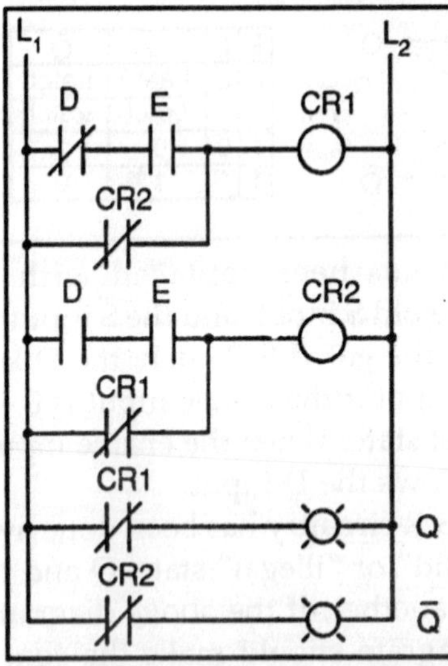

E	D	Q	$\bar{Q}$
0	0	Latch	Latch
0	1	Latch	Latch
1	0	0	1
1	1	1	0

An application for the D latch is a 1-bit memory circuit. You can "write" (store) a 0 or 1 bit in this latch circuit by making the enable input high (1) and setting D to whatever you want the stored bit to be. When the enable input is made low (0), the latch ignores the status of the D input and merrily holds the stored bit value, outputting at the stored value at Q, and its inverse on output not-Q.

Chapter 5

Sequential Logic Circuits

FLIP-FLOPS

So far, we've studied both S-R and D latch circuits with an enable inputs. The latch responds to the data inputs (S-R or D) only when the enable input is activated. In many digital applications, however, it is desirable to limit the responsiveness of a latch circuit to a very short period of time instead of the entire duration that the enabling input is activated. One method of enabling a multivibrator circuit is called *edge triggering,* where the circuit's data inputs have control only during the time that the enable input is *transitioning* from one state to another.

D FLIP FLOP

Let's compare timing diagrams for a normal D latch versus one that is edge-triggered:

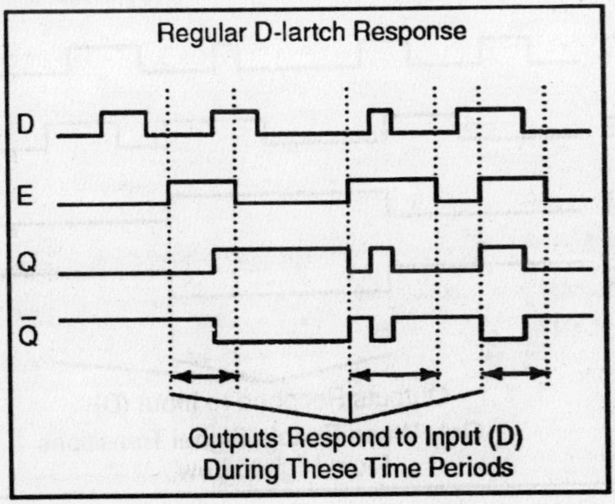

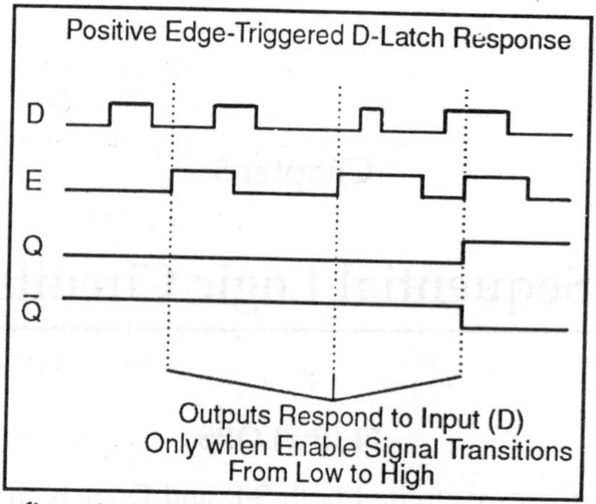

In the first timing diagram, the outputs respond to input D whenever the enable (E) input is high, for however long it remains high. When the enable signal falls back to a low state, the circuit remains latched. In the second timing diagram, we note a distinctly different response in the circuit output(s): it only responds to the D input during that brief moment of time when the enable signal *changes*, or *transitions*, from low to high. This is known as *positive* edge-triggering.

There is such a thing as *negative* edge triggering as well, and it produces the following response to the same input signals:

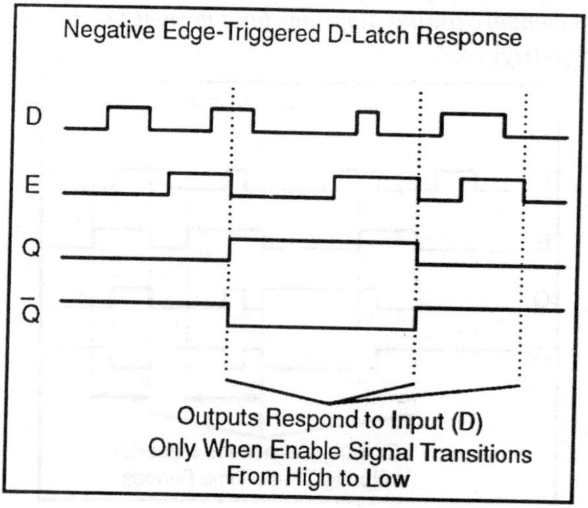

Whenever we enable a multivibrator circuit on the transitional edge of a square-wave enable signal, we call it a *flip-flop*instead of a *latch*.

Consequently, and edge-triggered S-R circuit is more properly known as an S-R flip-flop, and an edge-triggered D circuit as a D flip-flop.

The enable signal is renamed to be the *clock* signal. Also, we refer to the data inputs (S, R, and D, respectively) of these flip-flops as *synchronous* inputs, because they have effect only at the time of the clock pulse edge (transition), thereby synchronizing any output changes with that clock pulse, rather than at the whim of the data inputs.

But, how do we actually accomplish this edge-triggering? To create a "gated" S-R latch from a regular S-R latch is easy enough with a couple of AND gates, but how do we implement logic that only pays attention to the *rising or falling edge* of a changing digital signal?

What we need is a digital circuit that outputs a brief pulse whenever the input is activated for an arbitrary period of time, and we can use the output of this circuit to briefly enable the latch.

We're getting a little ahead of ourselves here, but this is actually a kind of monostable multivibrator, which for now we'll call a *pulse detector*.

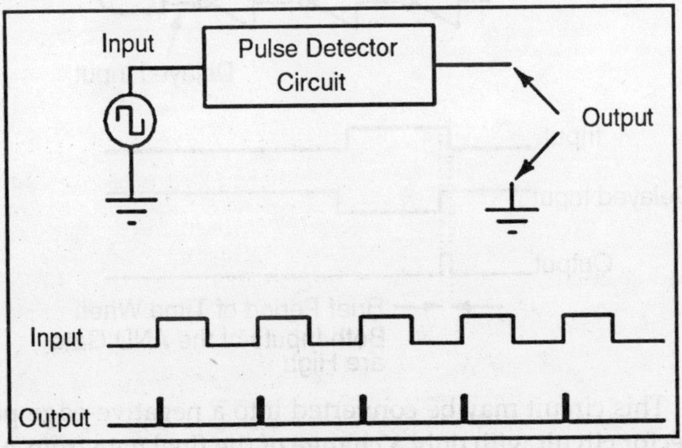

The duration of each output puise is set by components

in the pulse circuit itself. In ladder logic, this can be accomplished quite easily through the use of a time-delay relay with a very short delay time:

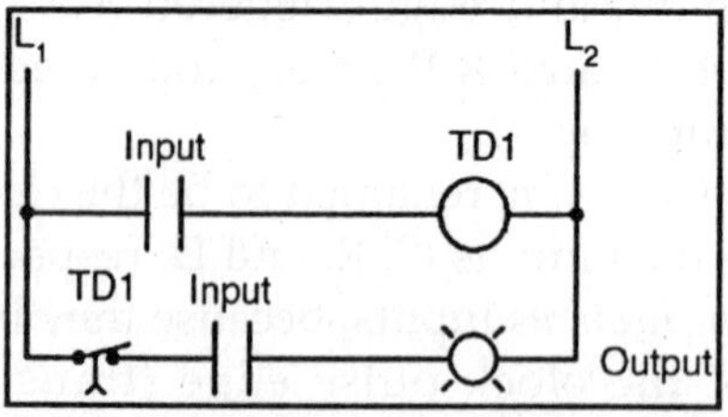

Implementing this timing function with semiconductor components is actually quite easy, as it exploits the inherent time delay within every logic gate (known as *propagation delay*).

What we do is take an input signal and split it up two ways, then place a gate or a series of gates in one of those signal paths just to delay it a bit, then have both the original signal and its delayed counterpart enter into a two-input gate that outputs a high signal for the brief moment of time that the delayed signal has not yet caught up to the low-to-high change in the non-delayed signal.

An example circuit for producing a clock pulse on a low-to-high input signal transition is shown here:

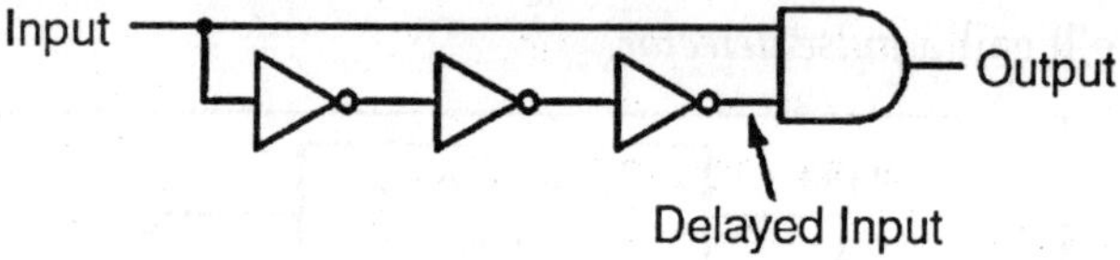

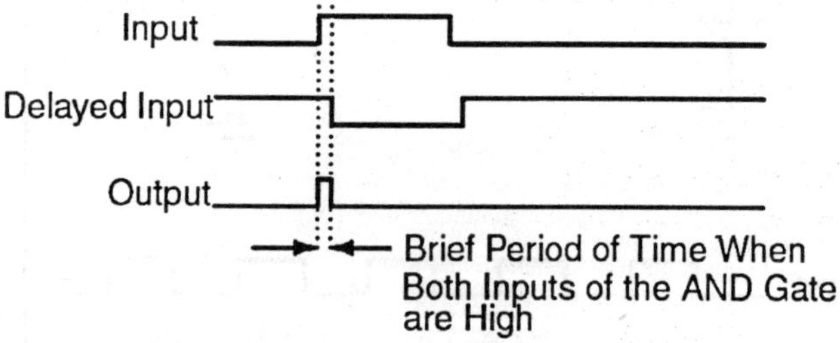

This circuit may be converted into a negative-edge pulse detector circuit with only a change of the final gate from AND to NOR:

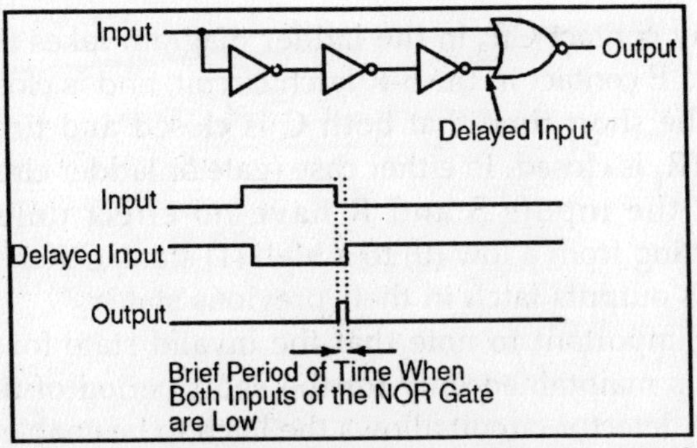

Now that we know how a pulse detector can be made, we can show it attached to the enable input of a latch to turn it into a flip-flop. In this case, the circuit is a S-R flip-flop:

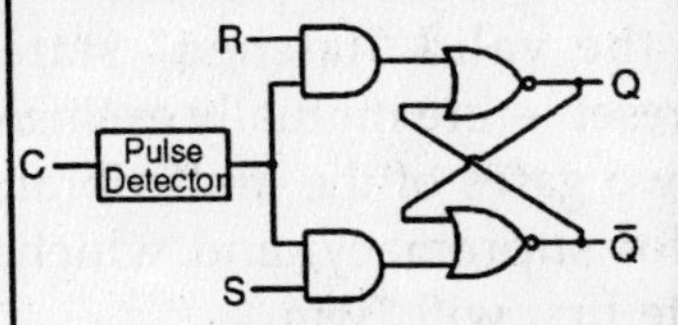

C	S	R	Q	$\bar{Q}$
↑	0	0	Latch	Latch
↑	0	1	0	1
↑	1	0	1	0
↑	1	1	0	0
X	0	0	Latch	Latch
X	0	1	Latch	Latch
X	1	0	Latch	Latch
X	1	1	Latch	Latch

Only when the clock signal (C) is transitioning from low to high is the circuit responsive to the S and R inputs. For any other condition of the clock signal ("x") the circuit will be latched.

A ladder logic version of the S-R flip-flop is shown here:

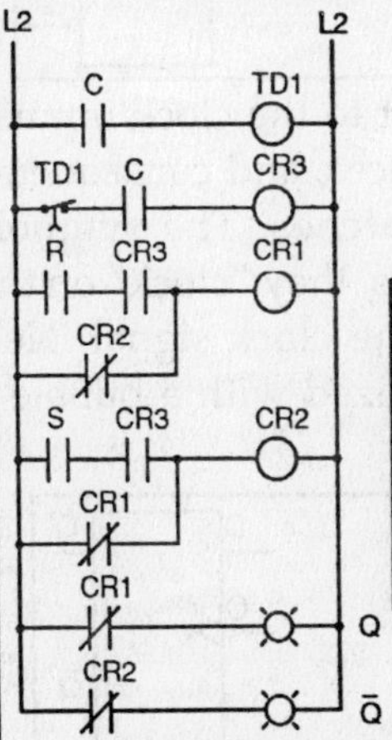

C	S	R	Q	$\bar{Q}$
↑	0	0	Latch	Latch
↑	0	1	0	1
↑	1	0	1	0
↑	1	1	0	0
X	0	0	Latch	Latch
X	0	1	Latch	Latch
X	1	0	Latch	Latch
X	1	1	Latch	Latch

Relay contact CR_3 in the ladder diagram takes the place of the old E contact in the S-R latch circuit, and is closed only during the short time that both C is closed and time-delay contact TR_1 is closed. In either case (gate or ladder circuit), we see that the inputs S and R have no effect unless C is transitioning from a low (0) to a high (1) state. Otherwise, the flip-flop's outputs latch in their previous states.

It is important to note that the invalid state for the S-R flip-flop is maintained only for the short period of time that the pulse detector circuit allows the latch to be enabled. After that brief time period has elapsed, the outputs will latch into either the set or the reset state.

Once again, the problem of a *race condition* manifests itself. With no enable signal, an invalid output state cannot be maintained. However, the valid "latched" states of the multivibrator — set and reset — are mutually exclusive to one another. Therefore, the two gates of the multivibrator circuit will "race" each other for supremacy, and whichever one attains a high output state first will "win."

The block symbols for flip-flops are slightly different from that of their respective latch counterparts:

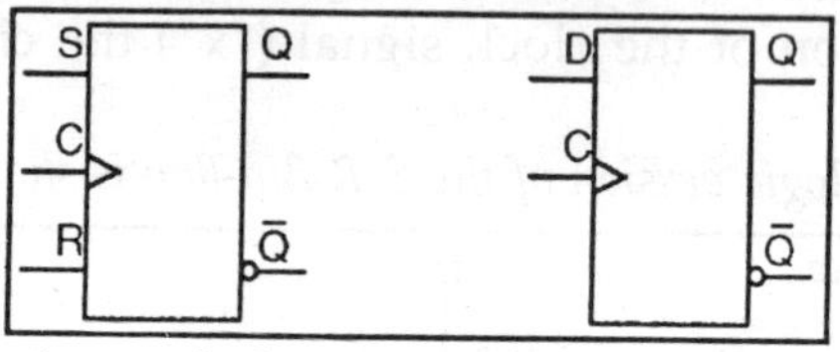

The triangle symbol next to the clock inputs tells us that these are edge-triggered devices, and consequently that these are flip-flops rather than latches. The symbols above are positive edge-triggered: that is, they "clock" on the rising edge (low-to-high transition) of the clock signal. Negative edge-triggered devices are symbolized with a bubble on the clock input line:

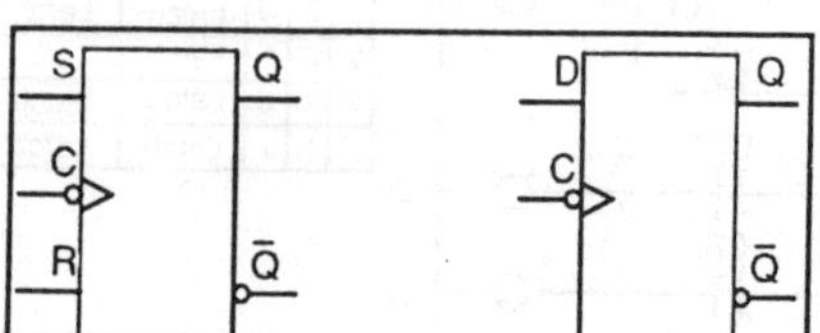

Both of the above flip-flops will "clock" on the falling edge (high-to-low transition) of the clock signal.

THE J-K FLIP-FLOP

Another variation on a theme of bistable multivibrators is the J-K flip-flop. Essentially, this is a modified version of an S-R flip-flop with no "invalid" or "illegal" output state. Look closely at the following diagram to see how this is accomplished:

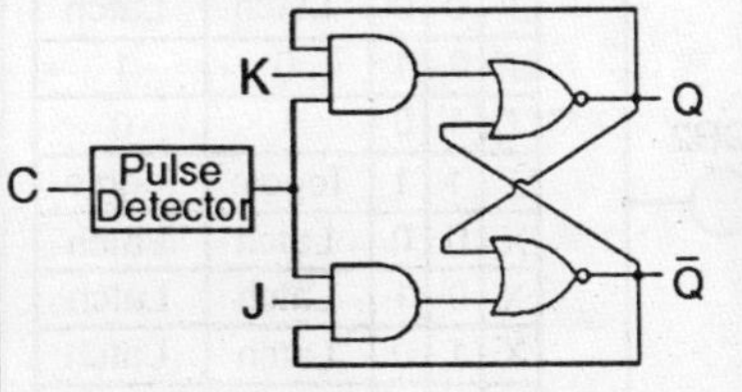

C	J	K	Q	$\bar{Q}$
↑	0	0	Latch	Latch
↑	0	1	0	1
↑	1	0	1	0
↑	1	1	Toggle	Toggle
X	0	0	Latch	Latch
X	0	1	Latch	Latch
X	1	0	Latch	Latch
X	1	1	Latch	Latch

What used to be the S and R inputs are now called the J and K inputs, respectively. The old two-input AND gates have been replaced with 3-input AND gates, and the third input of each gate receives feedback from the Q and not-Q outputs. What this does for us is permit the J input to have effect only when the circuit is reset, and permit the K input to have effect only when the circuit is set.

In other words, the two inputs are *interlocked*, to use a relay logic term, so that they cannot both be activated simultaneously. If the circuit is "set," the J input is inhibited by the 0 status of not-Q through the lower AND gate; if the circuit is "reset," the K input is inhibited by the 0 status of Q through the upper AND gate.

When both J and K inputs are 1, however, something unique happens. Because of the selective inhibiting action of those 3-input AND gates, a "set" state inhibits input J so that the flip-flop acts as if J=0 while K=1 when in fact both are 1. On the next clock pulse, the outputs will switch ("toggle") from set (Q=1 and not-Q=0) to reset (Q=0 and not-Q=1). Conversely, a "reset" state inhibits input K so that the flip-flop acts as if J=1 and K=0 when in fact both are 1. The next clock pulse toggles the circuit again from reset to set.

See if you can follow this logical sequence with the ladder logic equivalent of the J-K flip-flop:

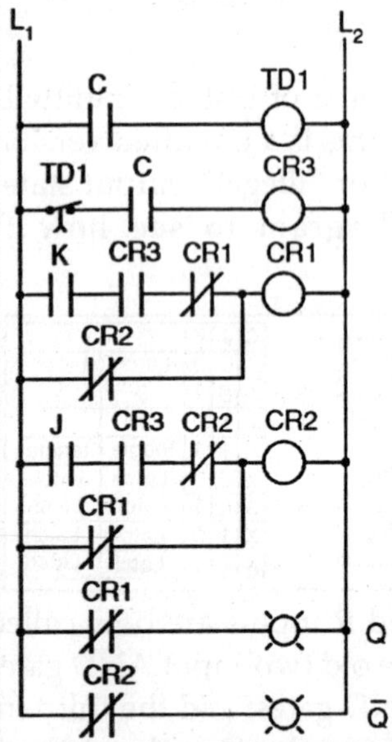

C	J	K	Q	$\overline{Q}$
↑	0	0	Latch	Latch
↑	0	1	0	1
↑	1	0	1	0
↑	1	1	Toggle	Toggle
X	0	0	Latch	Latch
X	0	1	Latch	Latch
X	1	0	Latch	Latch
X	1	1	Latch	Latch

The end result is that the S-R flip-flop's "invalid" state is eliminated (along with the race condition it engendered) and we get a useful feature as a bonus: the ability to toggle between the two (bistable) output states with every transition of the clock input signal.

There is no such thing as a J-K latch, only J-K flip-flops. Without the edge-triggering of the clock input, the circuit would continuously toggle between its two output states when both J and K were held high (1), making it an astable device instead of a bistable device in that circumstance.

If we want to preserve bistable operation for all combinations of input states, we *must* use edge-triggering so that it toggles only when we tell it to, one step (clock pulse) at a time.

The block symbol for a J-K flip-flop is a whole lot less frightening than its internal circuitry, and just like the S-R and D flip-flops, J-K flip-flops come in two clock varieties (negative and positive edge-triggered):

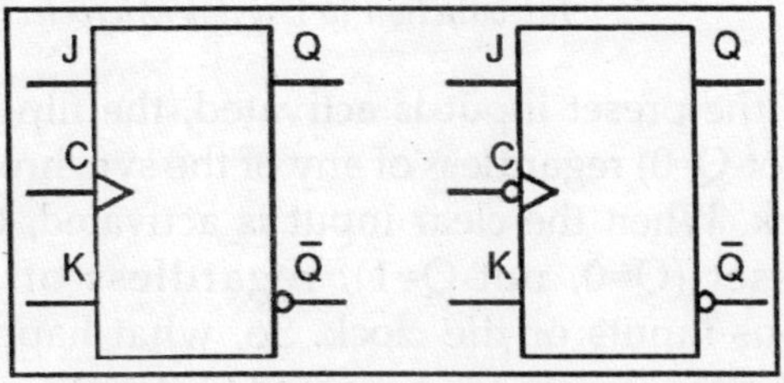

REVIEW

- A J-K flip-flop is nothing more than an S-R flip-flop with an added layer of feedback. This feedback selectively enables one of the two set/reset inputs so that they cannot both carry an active signal to the multivibrator circuit, thus eliminating the invalid condition.
- When both J and K inputs are activated, and the clock input is pulsed, the outputs (Q and not-Q) will swap states. That is, the circuit will toggle from a set state to a reset state, or visa-versa.

ASYNCHRONOUS FLIP-FLOP INPUTS

The normal data inputs to a flip flop (D, S and R, or J and K) are referred to as synchronous inputs because they have effect on the outputs (Q and not-Q) only in step, or in sync, with the clock signal transitions. These extra inputs that I now bring to your attention are called asynchronous because they can set or reset the flip-flop regardless of the status of the clock signal. Typically, they're called preset and clear:

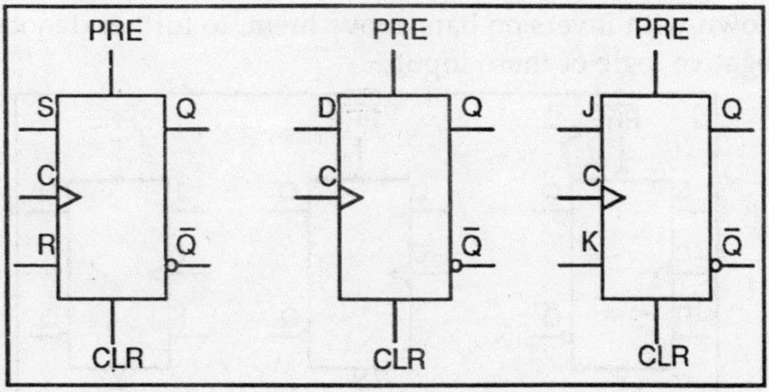

When the preset input is activated, the flip-flop will be set (Q=1, not-Q=0) regardless of any of the synchronous inputs or the clock. When the clear input is activated, the flip-flop will be reset (Q=0, not-Q=1), regardless of any of the synchronous inputs or the clock. So, what happens if both preset and clear inputs are activated? Surprise, surprise: we get an invalid state on the output, where Q and not-Q go to the same state, the same as our old friend, the S-R latch! Preset and clear inputs find use when multiple flip-flops are ganged together to perform a function on a multi-bit binary word, and a single line is needed to set or reset them all at once.

Asynchronous inputs, just like synchronous inputs, can be engineered to be active-high or active-low. If they're active-low, there will be an inverting bubble at that input lead on the block symbol, just like the negative edge-trigger clock inputs.

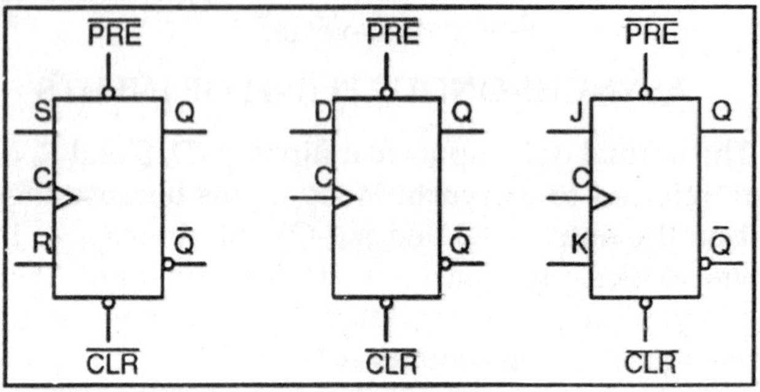

Sometimes the designations "PRE" and "CLR" will be shown with inversion bars above them, to further denote the negative logic of these inputs:

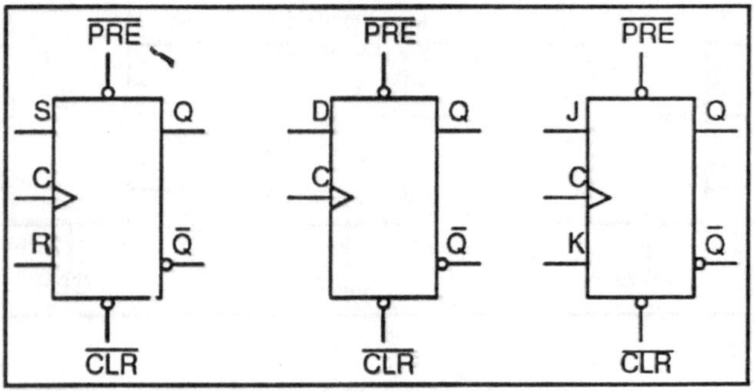

MONOSTABLE MULTIVIBRATORS

We've already seen one example of a monostable multivibrator in use: the pulse detector used within the circuitry of flip-flops, to enable the latch portion for a brief time when the clock input signal transitions from either low to high or high to low. The pulse detector is classified as a monostable multivibrator because it has only *one* stable state. By *stable,* I mean a state of output where the device is able to latch or hold to forever, without external prodding.

A latch or flip-flop, being a bistable device, can hold in either the "set" or "reset" state for an indefinite period of time. Once it's set or reset, it will continue to latch in that state unless prompted to change by an external input. A monostable device, on the other hand, is only able to hold in one particular state indefinitely. Its other state can only be held momentarily when triggered by an external input.

A mechanical analogy of a monostable device would be a momentary contact pushbutton switch, which spring-returns to its normal (stable) position when pressure is removed from its button actuator. Likewise, a standard wall (toggle) switch, such as the type used to turn lights on and off in a house, is a bistable device. It can latch in one of two modes: on or off.

All monostable multivibrators are *timed* devices. That is, their unstable output state will hold only for a certain minimum amount of time before returning to its stable state. With semiconductor monostable circuits, this timing function is typically accomplished through the use of resistors and capacitors, making use of the exponential charging rates of RC circuits.

A comparator is often used to compare the voltage across the charging (or discharging) capacitor with a steady reference voltage, and the on/off output of the comparator used for a logic signal. With ladder logic, time delays are accomplished with time-delay relays, which can be constructed with semiconductor/RC circuits like that just mentioned, or mechanical delay devices which impede the immediate motion of the relay's armature. Note the design and operation of the pulse detector circuit in ladder logic:

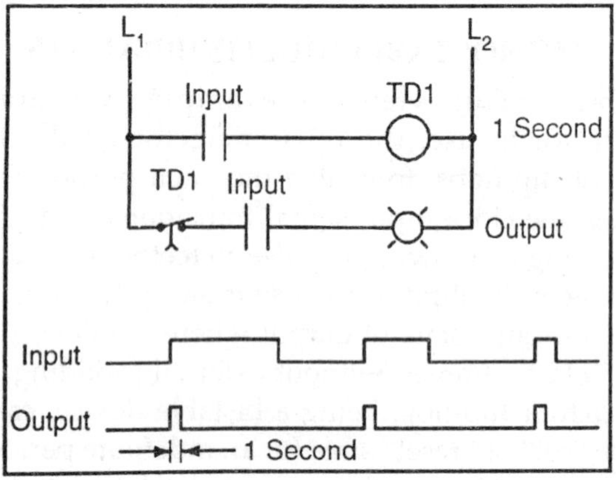

No matter how long the input signal stays high (1), the output remains high for just 1 second of time, then returns to its normal (stable) low state.

For some applications, it is necessary to have a monostable device that outputs a longer pulse than the input pulse which triggers it. Consider the following ladder logic circuit:

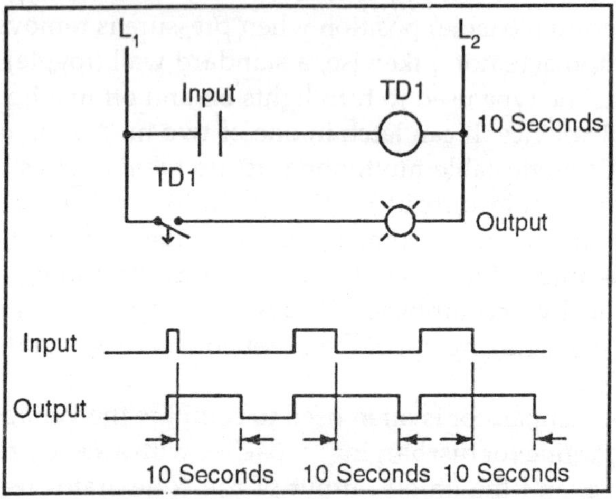

When the input contact closes, TD_1 contact immediately closes, and stays closed for 10 seconds after the input contact opens.

No matter how short the input pulse is, the output stays high (1) for exactly 10 seconds after the input drops low again.

This kind of monostable multivibrator is called a *one-shot*. More specifically, it is a *retriggerable* one-shot, because the timing begins after the input drops to a low state, meaning that multiple input pulses within 10 seconds of each other will maintain a continuous high output:

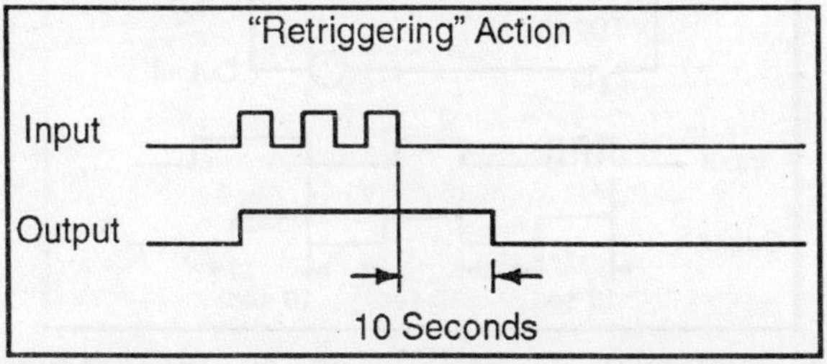

One application for a retriggerable one-shot is that of a single mechanical contact debouncer. As you can see from the above timing diagram, the output will remain high despite "bouncing" of the input signal from a mechanical switch. Of course, in a real-life switch debouncer circuit, you'd probably want to use a time delay of much shorter duration than 10 seconds, as you only need to "debounce" pulses that are in the millisecond range.

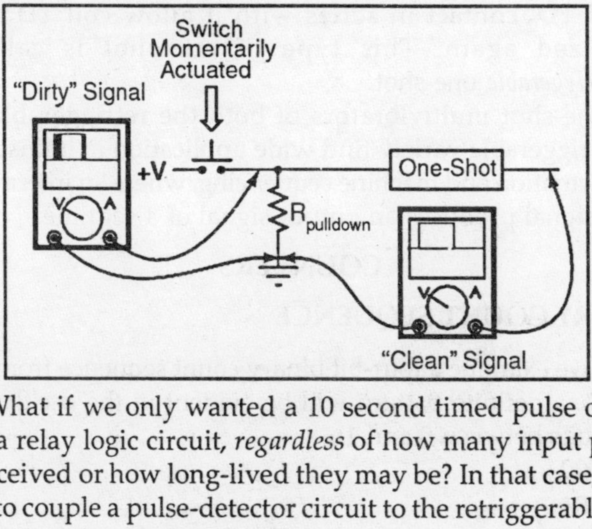

What if we only wanted a 10 second timed pulse output from a relay logic circuit, *regardless* of how many input pulses we received or how long-lived they may be? In that case, we'd have to couple a pulse-detector circuit to the retriggerable one-shot time delay circuit, like this:

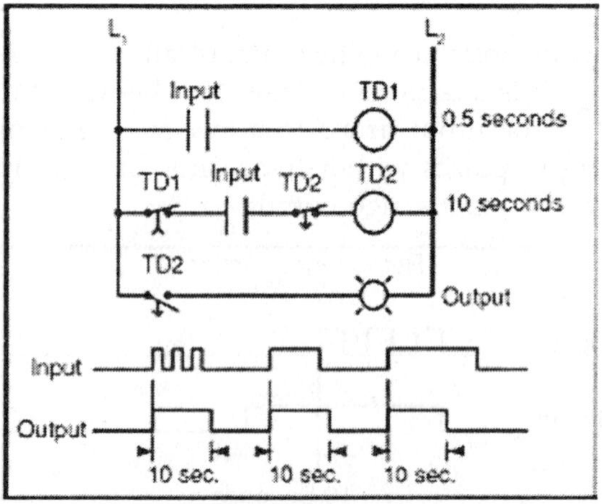

Time delay relay TD_1 provides an "on" pulse to time delay relay coil TD_2 for an arbitrarily short moment (in this circuit, for at least 0.5 second each time the input contact is actuated). As soon as TD_2 is energized, the normally-closed, timed-closed TD_2 contact in series with it prevents coil TD_2 from being re-energized as long as it's timing out (10 seconds). This effectively makes it unresponsive to any more actuations of the input switch during that 10 second period.

Only after TD_2 times out does the normally-closed, timed-closed TD_2 contact in series with it allow coil TD_2 to be energized again. This type of one-shot is called a *nonretriggerable* one-shot.

One-shot multivibrators of both the retriggerable and nonretriggerable variety find wide application in industry for siren actuation and machine sequencing, where an intermittent input signal produces an output signal of a set time.

COUNTERS

BINARY COUNT SEQUENCE

If we examine a four-bit binary count sequence from 0000 to 1111, a definite pattern will be evident in the"oscillations" of the bits between 0 and 1:

0000
0001

0010
0011
0100
0101
0110
0111
1000
1001
1010
1011
1100
1101
1110
1111

Note how the least significant bit (LSB) toggles between 0 and 1 for every step in the count sequence, while each succeeding bit toggles at one-half the frequency of the one before it.

The most significant bit (MSB) only toggles once during the entire sixteen-step count sequence: at the transition between 7 (0111) and 8 (1000).

If we wanted to design a digital circuit to"count" in four-bit binary, all we would have to do is design a series of frequency divider circuits, each circuit dividing the frequency of a square-wave pulse by a factor of 2:

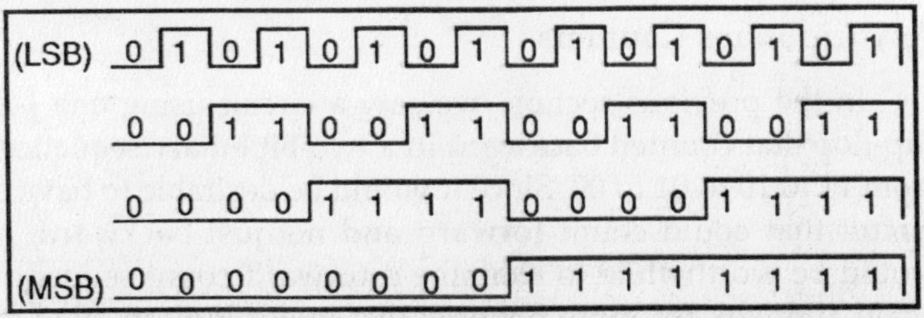

J-K flip-flops are ideally suited for this task, because they have the ability to"toggle" their output state at the command of a clock pulse when both J and K inputs are made"high" (1):

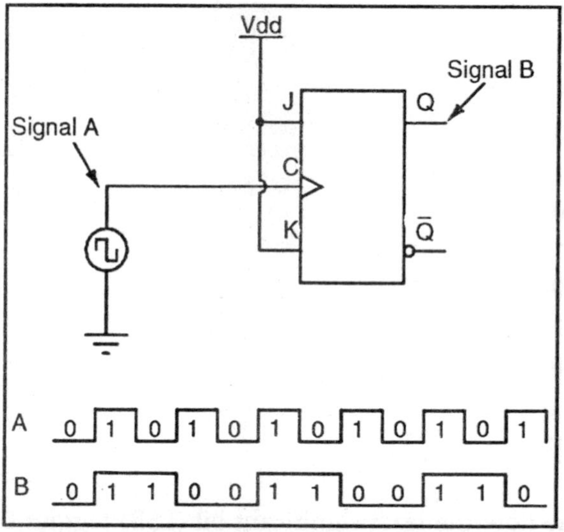

If we consider the two signals (A and B) in this circuit to represent two bits of a binary number, signal A being the LSB and signal B being the MSB, we see that the count sequence is backward: from 11 to 10 to 01 to 00 and back again to 11. Although it might not be counting in the direction we might have assumed, at least it counts! The following sections explore different types of counter circuits, all made with J-K flip-flops, and all based on the exploitation of that flip-flop's toggle mode of operation.

TYPES OF COUNTERS

Asynchronous Counters

In the previous section, we saw a circuit using one J-K flip-flop that counted backward in a two-bit binary sequence, from 11 to 10 to 01 to 00. Since it would be desirable to have a circuit that could count forward and not just backward, it would be worthwhile to examine a forward count sequence again and look for more patterns that might indicate how to build such a circuit.

Since we know that binary count sequences follow a pattern of octave (factor of 2) frequency division, and that J-K flip-flop multivibrators set up for the"toggle" mode are capable of performing this type of frequency division, we can envision a circuit made up of several J-K flip-flops, cascaded to produce four bits of output.

The main problem facing us is to determine how to connect these flip-flops together so that they toggle at the right times to produce the proper binary sequence. Examine the following binary count sequence, paying attention to patterns preceding the"toggling" of a bit between 0 and 1:

0000
0001
0010
0011
0100
0101
0110
0111
1000
1001
1010
1011
1100
1101
1110
1111

Note that each bit in this four-bit sequence toggles when the bit before it (the bit having a lesser significance, or place-weight), toggles in a particular direction: from 1 to 0. Small arrows indicate those points in the sequence where a bit toggles, the head of the arrow pointing to the previous bit transitioning from a"high" (1) state to a"low" (0) state:

0000
0001
 →
0010
0011
→→

0100
0101
→
0110
0111
→→→
1000
1001
→
1010
1011
→→
1100
1101
→
1110
1111

Starting with four J-K flip-flops connected in such a way to always be in the"toggle" mode, we need to determine how to connect the clock inputs in such a way so that each succeeding bit toggles when the bit before it transitions from 1 to 0. The Q outputs of each flip-flop will serve as the respective binary bits of the final, four-bit count:

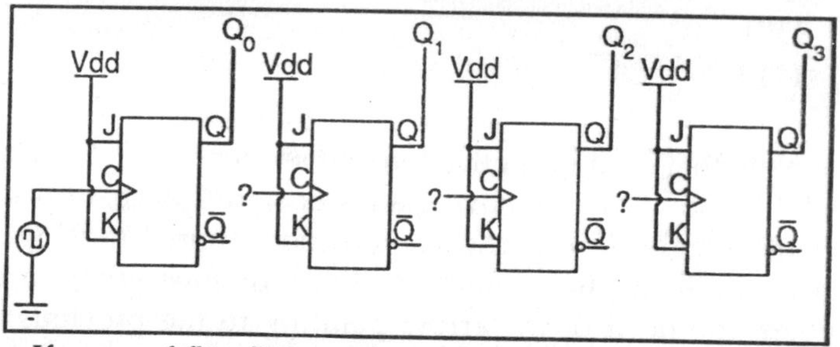

If we used flip-flops with negative-edge triggering (bubble symbols on the clock inputs), we could simply connect the clock input of each flip-flop to the Q output of the flip-flop before it, so that when the bit before it changes from a 1 to a 0, the"falling edge" of that signal would"clock" the next flip-flop to toggle the next bit:

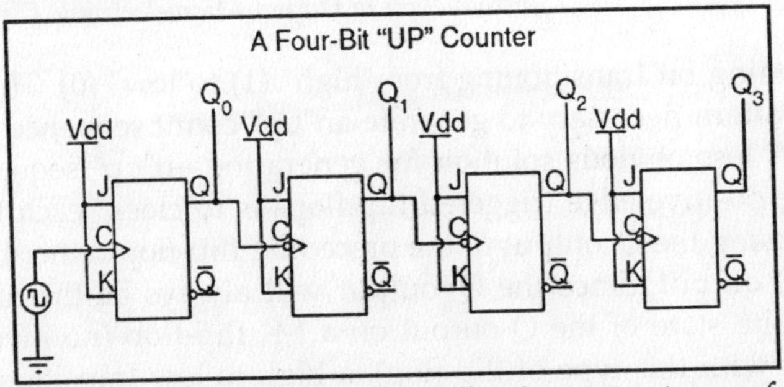

This circuit would yield the following output waveforms, when"clocked" by a repetitive source of pulses from an oscillator:

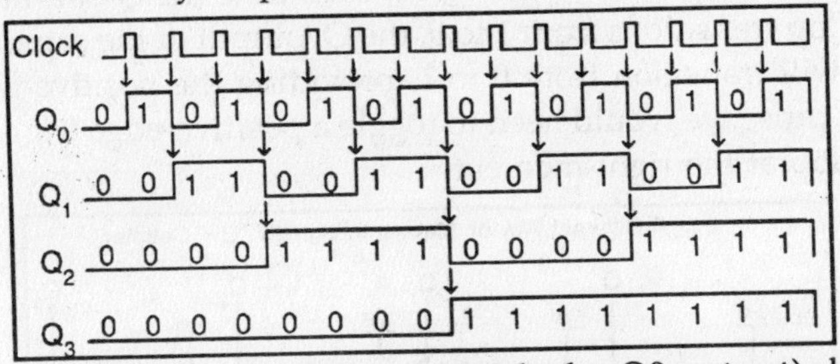

The first flip-flop (the one with the Q0 output), has a positive-edge triggered clock input, so it toggles with each rising edge of the clock signal. Notice how the clock signal in this example has a duty cycle less than 50%. I've shown the signal in this manner for the purpose of demonstrating how the clock signal need not be symmetrical to obtain reliable,"clean" output bits in our four-bit binary sequence. In the very first flip-flop circuit shown in this chapter, I used the clock signal itself as one of the output bits.

This is a bad practice in counter design, though, because it necessitates the use of a square wave signal with a 50% duty cycle ("high" time ="low" time) in order to obtain a count sequence where each and every step pauses for the same amount of time. Using one J-K flip-flop for each output bit, however, relieves us of the necessity of having a symmetrical clock signal, allowing the use of practically any variety of high/low waveform to increment the count sequence.

As indicated by all the other arrows in the pulse diagram, each succeeding output bit is toggled by the action of the preceding bit transitioning from"high" (1) to"low" (0). This is the pattern necessary to generate an"up" count sequence.

A less obvious solution for generating an"up" sequence using positive-edge triggered flip-flops is to"clock" each flip-flop using the Q' output of the preceding flip-flop rather than the Q output. Since the Q' output will always be the exact opposite state of the Q output on a J-K flip-flop (no invalid states with this type of flip-flop), a high-to-low transition on the Q output will be accompanied by a low-to-high transition on the Q' output. In other words, each time the Q output of a flip-flop transitions from 1 to 0, the Q' output of the same flip-flop will transition from 0 to 1, providing the positive-going clock pulse we would need to toggle a positive-edge triggered flip-flop at the right moment:

A Different Way of Making a Four-BIT "UP" Counter

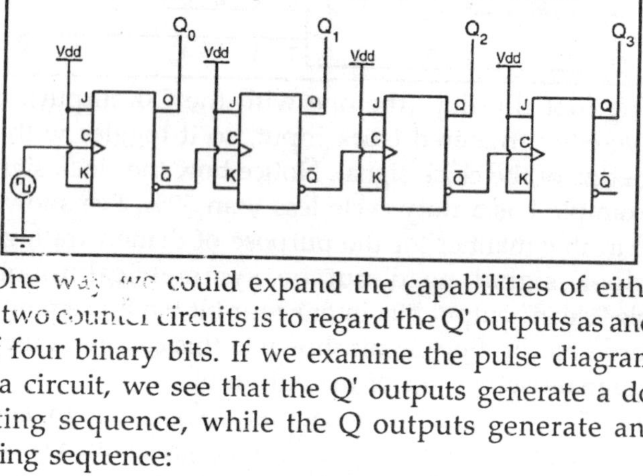

One way we could expand the capabilities of either of these two counter circuits is to regard the Q' outputs as another set of four binary bits. If we examine the pulse diagram for such a circuit, we see that the Q' outputs generate a down-counting sequence, while the Q outputs generate an up-counting sequence:

A Simultaneous "Up" and "Down" Counter

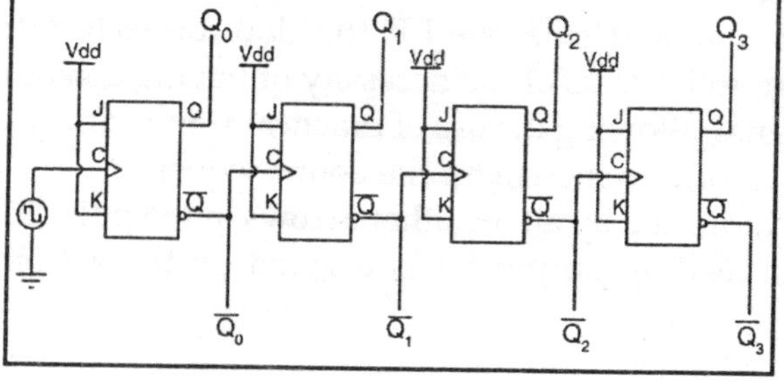

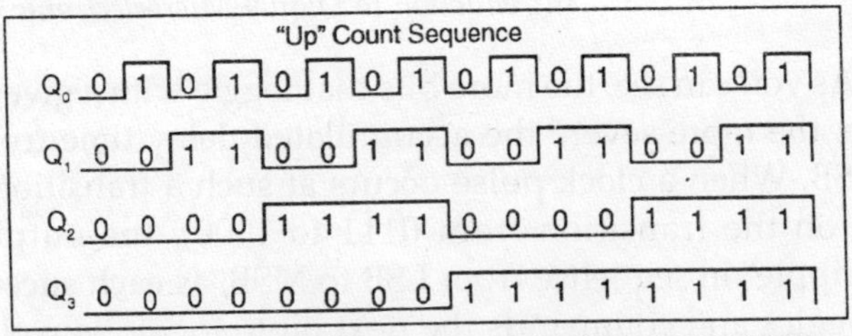

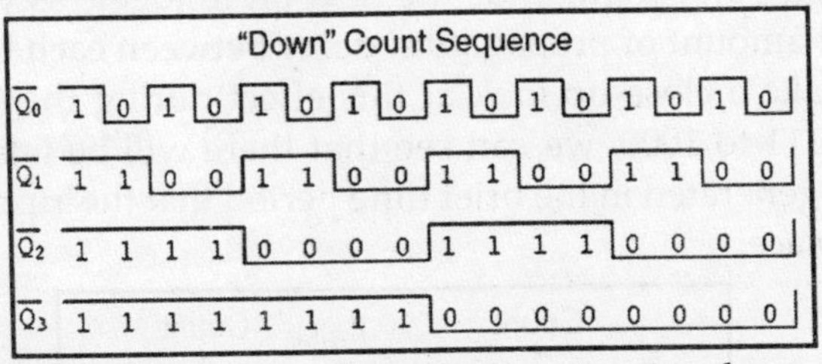

Unfortunately, all of the counter circuits shown thusfar share a common problem: the ripple effect. This effect is seen in certain types of binary adder and data conversion circuits, and is due to accumulative propagation delays between cascaded gates. When the Q output of a flip-flop transitions from 1 to 0, it commands the next flip-flop to toggle. If the next flip-flop toggle is a transition from 1 to 0, it will command the flip-flop after it to toggle as well, and so on.

However, since there is always some small amount of propagation delay between the command to toggle (the clock pulse) and the actual toggle response (Q and Q' outputs changing states), any subsequent flip-flops to be toggled will toggle some time after the first flip-flop has toggled. Thus, when multiple bits toggle in a binary count sequence, they will not all toggle at exactly the same time:

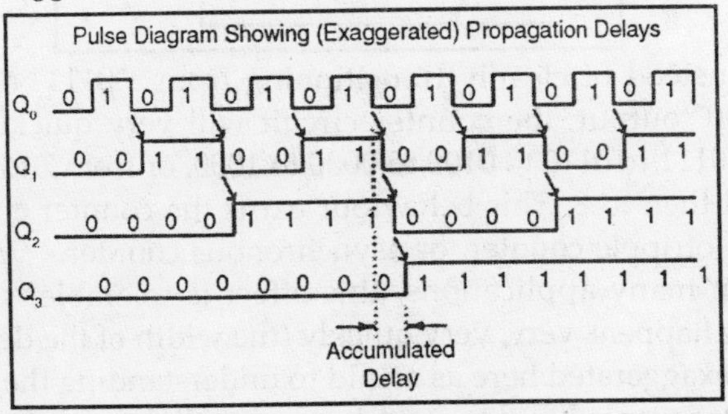

As you can see, the more bits that toggle with a given clock pulse, the more severe the accumulated delay time from LSB to MSB. When a clock pulse occurs at such a transition point (say, on the transition from 0111 to 1000), the output bits will"ripple" in sequence from LSB to MSB, as each succeeding bit toggles and commands the next bit to toggle as well, with a small amount of propagation delay between each bit toggle. If we take a close-up look at this effect during the transition from 0111 to 1000, we can see that there will be false output counts generated in the brief time period that the"ripple" effect takes place:

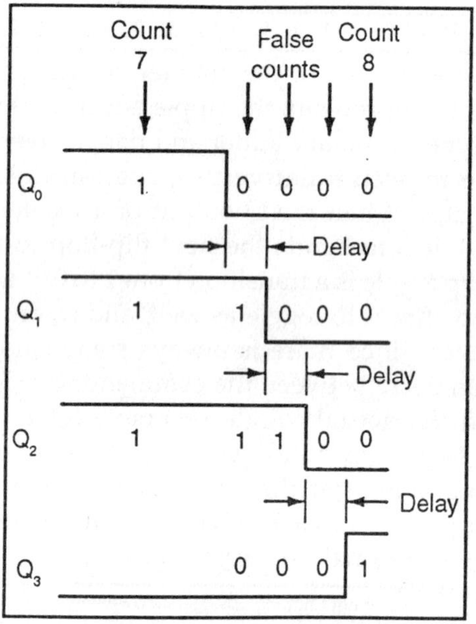

Instead of cleanly transitioning from a"0111" output to a"1000" output, the counter circuit will very quickly ripple from 0111 to 0110 to 0100 to 0000 to 1000, or from 7 to 6 to 4 to 0 and then to 8. This behaviour earns the counter circuit the name ofripple counter, or asynchronous counter.

In many applications, this effect is tolerable, since the ripple happens very, very quickly (the width of the delays has been exaggerated here as an aid to understanding the effects). If all we wanted to do was drive a set of light-emitting diodes

(LEDs) with the counter's outputs, for example, this brief ripple would be of no consequence at all.

However, if we wished to use this counter to drive the"select" inputs of a multiplexer, index a memory pointer in a microprocessor (computer) circuit, or perform some other task where false outputs could cause spurious errors, it would not be acceptable. There is a way to use this type of counter circuit in applications sensitive to false, ripple-generated outputs, and it involves a principle known as strobing.

Most decoder and multiplexer circuits are equipped with at least one input called the"enable." The output(s) of such a circuit will be active only when the enable input is made active. We can use this enable input to strobe the circuit receiving the ripple counter's output so that it is disabled (and thus not responding to the counter output) during the brief period of time in which the counter outputs might be rippling, and enabled only when sufficient time has passed since the last clock pulse that all rippling will have ceased. In most cases, the strobing signal can be the same clock pulse that drives the counter circuit:

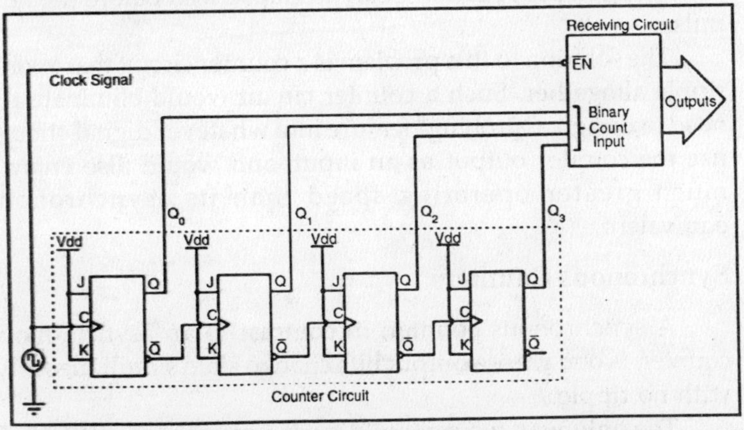

With an active-low Enable input, the receiving circuit will respond to the binary count of the four-bit counter circuit only when the clock signal is"low." As soon as the clock pulse goes"high," the receiving circuit stops responding to the counter circuit's output. Since the counter circuit is positive-edge triggered (as determined by the first flip-flop clock input), all the counting action takes place on the low-to-high transition of the clock signal, meaning that the receiving circuit will become disabled just before any toggling occurs on the counter circuit's four output bits.

The receiving circuit will not become enabled until the clock signal returns to a low state, which should be a long enough time after all rippling has ceased to be"safe" to allow the new count to have effect on the receiving circuit. The crucial parameter here is the clock signal's"high" time: it must be at least as long as the maximum expected ripple period of the counter circuit. If not, the clock signal will prematurely enable the receiving circuit, while some rippling is still taking place.

Another disadvantage of the asynchronous, or ripple, counter circuit is limited speed. While all gate circuits are limited in terms of maximum signal frequency, the design of asynchronous counter circuits compounds this problem by making propagation delays additive.

Thus, even if strobing is used in the receiving circuit, an asynchronous counter circuit cannot be clocked at any frequency higher than that which allows the greatest possible accumulated propagation delay to elapse well before the next pulse.

The solution to this problem is a counter circuit that avoids ripple altogether. Such a counter circuit would eliminate the need to design a"strobing" feature into whatever digital circuits use the counter output as an input, and would also enjoy a much greater operating speed than its asynchronous equivalent.

Synchronous Counters

A synchronous counter, in contrast to an asynchronous counter, is one whose output bits change state simultaneously, with no ripple.

The only way we can build such a counter circuit from J-K flip-flops is to connect all the clock inputs together, so that each and every flip-flop receives the exact same clock pulse at the exact same time:

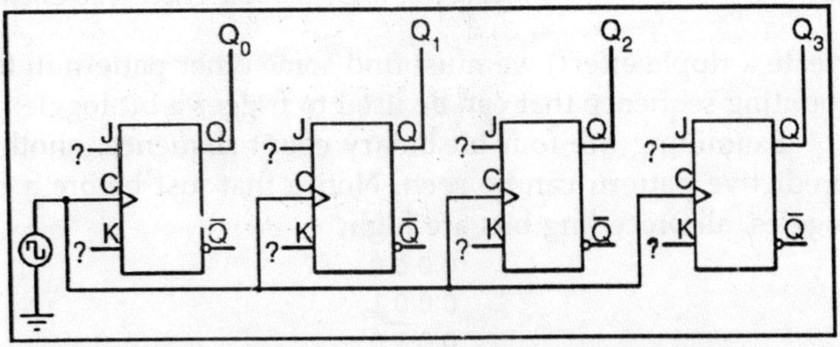

Now, the question is, what do we do with the J and K inputs? We know that we still have to maintain the same divide-by-two frequency pattern in order to count in a binary sequence, and that this pattern is best achieved utilizing the"toggle" mode of the flip-flop, so the fact that the J and K inputs must both be (at times)"high" is clear.

However, if we simply connect all the J and K inputs to the positive rail of the power supply as we did in the asynchronous circuit, this would clearly not work because all the flip-flops would toggle at the same time: with each and every clock pulse!

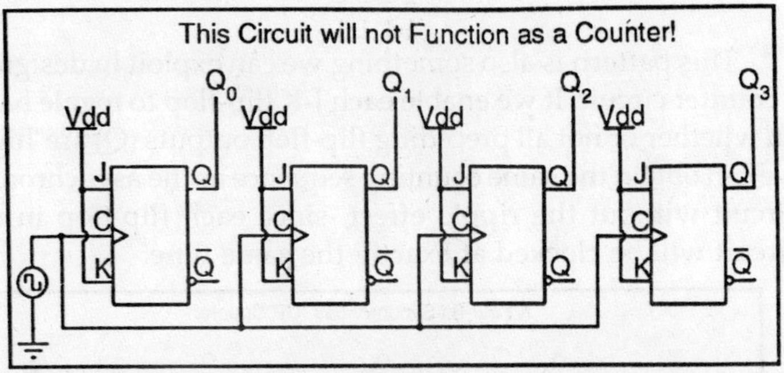

Let's examine the four-bit binary counting sequence again, and see if there are any other patterns that predict the toggling of a bit.

Asynchronous counter circuit design is based on the fact that each bit toggle happens at the same time that the preceding bit toggles from a"high" to a"low" (from 1 to 0). Since we cannot clock the toggling of a bit based on the toggling of a previous bit in a synchronous counter circuit (to do so would create a ripple effect) we must find some other pattern in the counting sequence that can be used to trigger a bit toggle:

Examining the four-bit binary count sequence, another predictive pattern can be seen. Notice that just before a bit toggles, all preceding bits are"high:"

0000
0001
0010
0011
0100
0101
0110
0111
1000
1001
1010
1011
1100
1101
1110
1111

This pattern is also something we can exploit in designing a counter circuit. If we enable each J-K flip-flop to toggle based on whether or not all preceding flip-flop outputs (Q) are"high," we can obtain the same counting sequence as the asynchronous circuit without the ripple effect, since each flip-flop in this circuit will be clocked at exactly the same time:

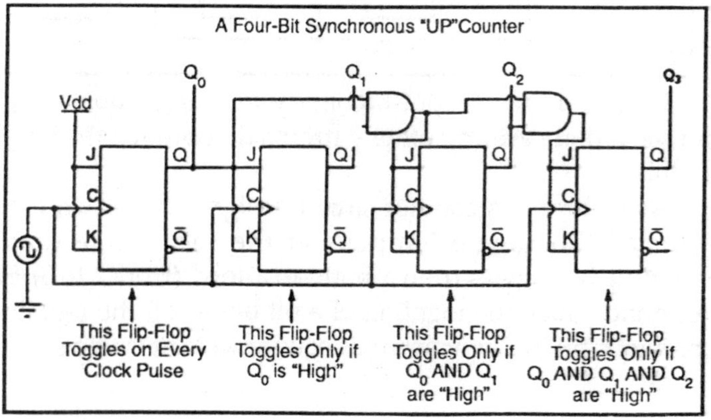

A Four-Bit Synchronous "UP"Counter

The result is a four-bit synchronous"up" counter. Each of the higher-order flip-flops are made ready to toggle (both J and K inputs"high") if the Q outputs of all previous flip-flops are"high." Otherwise, the J and K inputs for that flip-flop will both be"low," placing it into the"latch" mode where it will maintain its present output state at the next clock pulse. Since the first (LSB) flip-flop needs to toggle at every clock pulse, its J and K inputs are connected to Vcc or Vdd, where they will be"high" all the time.

The next flip-flop need only"recognize" that the first flip-flop's Q output is high to be made ready to toggle, so no AND gate is needed. However, the remaining flip-flops should be made ready to toggle only when all lower-order output bits are"high," thus the need for AND gates.

To make a synchronous"down" counter, we need to build the circuit to recognize the appropriate bit patterns predicting each toggle state while counting down. Not surprisingly, when we examine the four-bit binary count sequence, we see that all preceding bits are"low" prior to a toggle (following the sequence from bottom to top):

```
0000
0001
0010
0011
0100
0101
0110
0111
1000
1001
1010
1011
1100
1101
1110
1111
```

Since each J-K flip-flop comes equipped with a Q' output

as well as a Q output, we can use the Q' outputs to enable the toggle mode on each succeeding flip-flop, being that each Q' will be"high" every time that the respective Q is"low:"

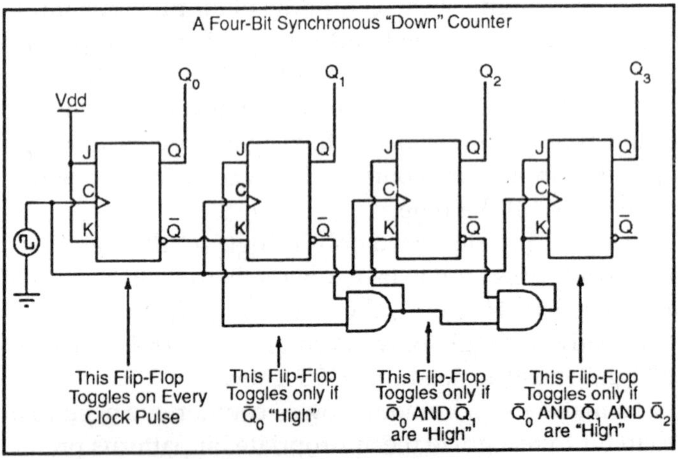

Taking this idea one step further, we can build a counter circuit with selectable between"up" and"down" count modes by having dual lines of AND gates detecting the appropriate bit conditions for an"up" and a"down" counting sequence, respectively, then use OR gates to combine the AND gate outputs to the J and K inputs of each succeeding flip-flop:

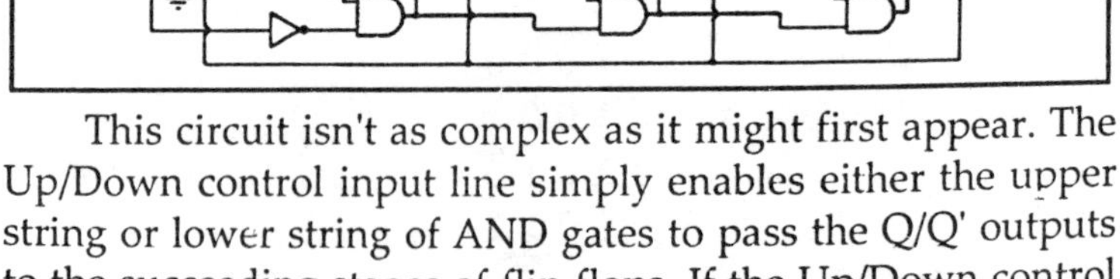

This circuit isn't as complex as it might first appear. The Up/Down control input line simply enables either the upper string or lower string of AND gates to pass the Q/Q' outputs to the succeeding stages of flip-flops. If the Up/Down control line is"high," the top AND gates become enabled, and the circuit functions exactly the same as the first ("up") synchronous counter circuit shown in this section. If the Up/

Down control line is made"low," the bottom AND gates become enabled, and the circuit functions identically to the second ("down" counter) circuit shown in this section.

To illustrate, here is a diagram showing the circuit in the"up" counting mode (all disabled circuitry shown in grey rather than black):

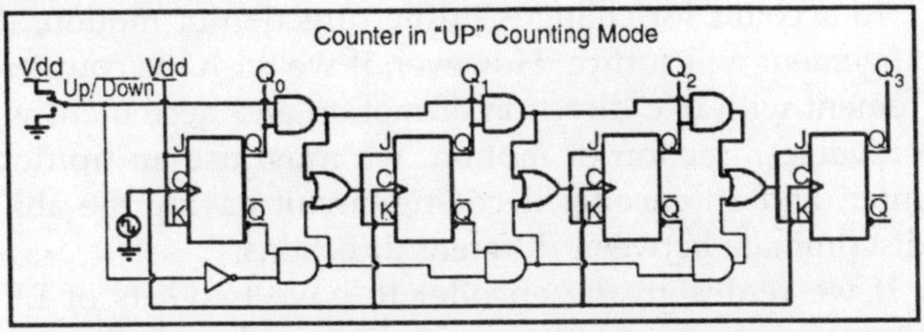

Here, shown in the"down" counting mode, with the same grey colouring representing disabled circuitry:

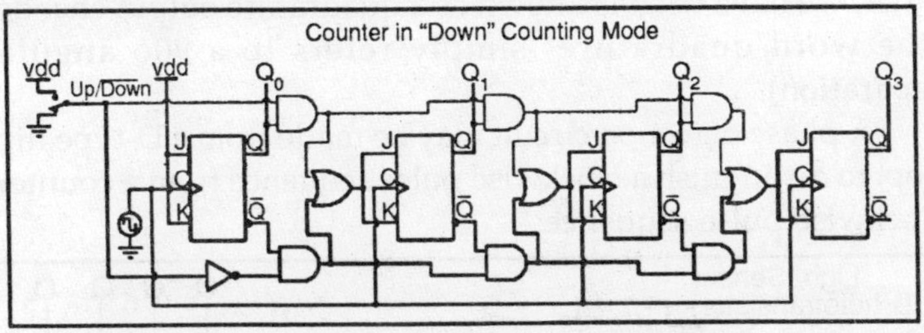

Up/down counter circuits are very useful devices. A common application is in machine motion control, where devices calledrotary shaft encoders convert mechanical rotation into a series of electrical pulses, these pulses"clocking" a counter circuit to track total motion:

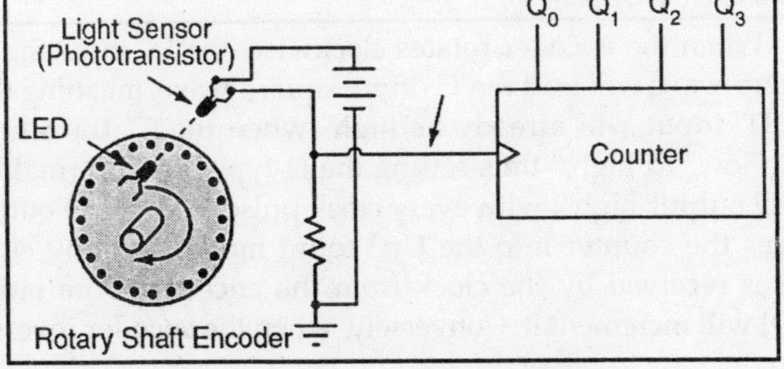

As the machine moves, it turns the encoder shaft, making and breaking the light beam between LED and phototransistor, thereby generating clock pulses to increment the counter circuit. Thus, the counter integrates, or accumulates, total motion of the shaft, serving as an electronic indication of how far the machine has moved.

If all we care about is tracking total motion, and do not care to account for changes in the direction of motion, this arrangement will suffice. However, if we wish the counter to increment with one direction of motion and decrement with the reverse direction of motion, we must use an up/down counter, and an encoder/decoding circuit having the ability to discriminate between different directions.

If we re-design the encoder to have two sets of LED/ phototransistor pairs, those pairs aligned such that their square-wave output signals are 90o out of phase with each other, we have what is known as a quadrature output encoder (the word"quadrature" simply refers to a 90o angular separation).

A phase detection circuit may be made from a D-type flip-flop, to distinguish a clockwise pulse sequence from a counter-clockwise pulse sequence:

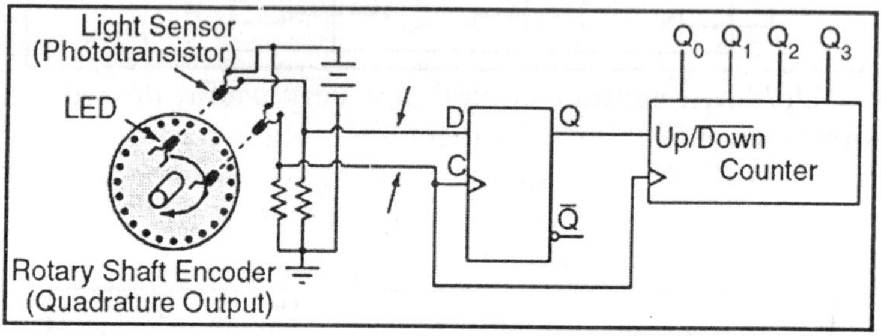

When the encoder rotates clockwise, the"D" input signal square-wave will lead the"C" input square-wave, meaning that the"D" input will already be"high" when the"C" transitions from"low" to"high," thus setting the D-type flip-flop (making the Q output"high") with every clock pulse. A"high" Q output places the counter into the"Up" count mode, and any clock pulses received by the clock from the encoder (from either LED) will increment it. Conversely, when the encoder reverses rotation, the"D" input will lag behind the"C" input waveform, meaning that it will be"low" when the"C" waveform transitions from"low" to"high," forcing the D-type flip-flop into the reset state (making the Q output"low") with every clock pulse. This"low" signal commands the counter circuit to decrement with every clock pulse from the encoder.

This circuit, or something very much like it, is at the heart of every position-measuring circuit based on a pulse encoder sensor. Such applications are very common in robotics, CNC machine tool control, and other applications involving the measurement of reversible, mechanical motion.

Chapter 6

Combinational Logic Circuits

DIGITAL COMPUTING

A BINARY ADDER

Suppose we wanted to build a device that could add two binary bits together. Such a device is known as a half-adder, and its gate circuit looks like this:

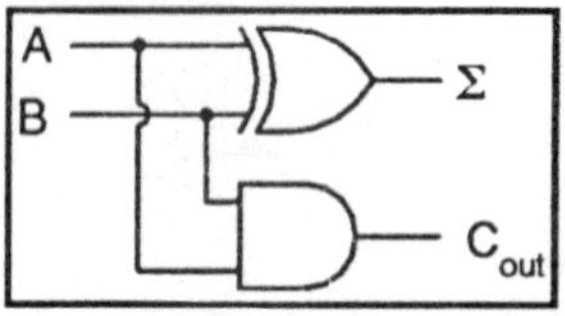

The Σ symbol represents the "sum" output of the half-adder, the sum's least significant bit (LSB). C_{out} represents the "carry" output of the half-adder, the sum's most significant bit (MSB).

If we were to implement this same function in ladder (relay) logic, it would look like this:

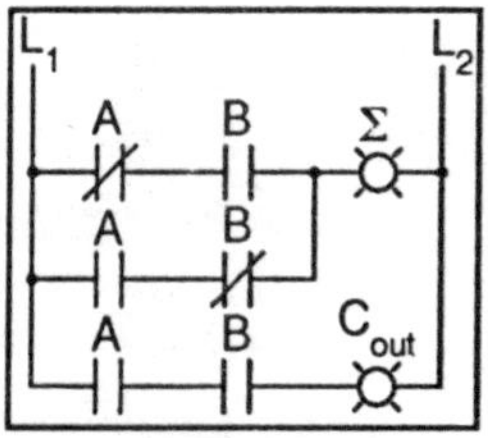

Either circuit is capable of adding two binary digits together. The mathematical "rules" of how to add bits together are intrinsic to the hard-wired logic of the circuits. If we wanted to perform a different arithmetic operation with binary bits, such as multiplication, we would have to construct another circuit. The above circuit designs will only perform one function: add two binary bits together. To make them do something else would take re-wiring, and perhaps different componentry.

In this sense, digital arithmetic circuits aren't much different from analog arithmetic (operational amplifier) circuits: they do exactly what they're wired to do, no more and no less. We are not, however, restricted to designing digital computer circuits in this manner. It is possible to embed the mathematical "rules" for any arithmetic operation in the form of digital data rather than in hard-wired connections between gates. The result is unparalleled flexibility in operation, giving rise to a whole new kind of digital device: the *programmable computer*.

While this chapter is by no means exhaustive, it provides what I believe is a unique and interesting look at the nature of programmable computer devices, starting with two devices often overlooked in introductory textbooks: *look-up table memories* and *finite-state machines*.

LOOK-UP TABLES

Having learned about digital memory devices in the last chapter, we know that it is possible to store binary data within solid-state devices. Those storage "cells" within solid-state memory devices are easily addressed by driving the "address" lines of the device with the proper binary value(s). Suppose we had a ROM memory circuit written, or programmed, with certain data, such that the address lines of the ROM served as inputs and the data lines of the ROM served as outputs, generating the characteristic response of a particular logic function. Theoretically, we could programme this ROM chip to emulate whatever logic function we wanted without having to alter any wire connections or gates.

Consider the following example of a 4 × 2 bit ROM memory (a very small memory!) programmed with the functionality of a half adder:

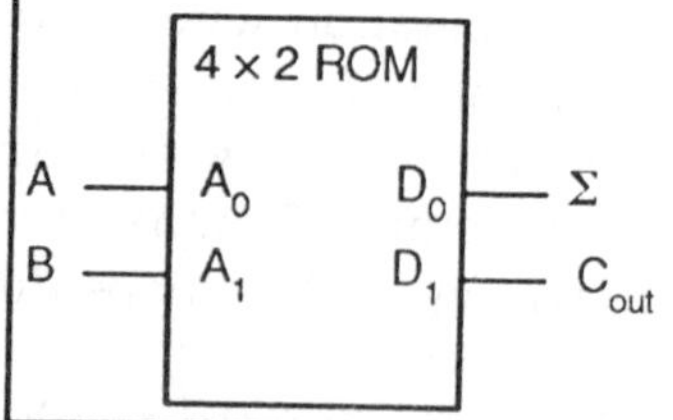

Address		Data	
A	B	C_{out}	Σ
0	0	0	0
0	1	0	1
1	0	0	1
1	1	1	0

If this ROM has been written with the above data (representing a half-adder's truth table), driving the A and B address inputs will cause the respective memory cells in the ROM chip to be enabled, thus outputting the corresponding data as the Ó (Sum) and C_{out} bits.

Unlike the half-adder circuit built of gates or relays, this device can be set up to perform any logic function at all with two inputs and two outputs, not just the half-adder function. To change the logic function, all we would need to do is write a different table of data to another ROM chip. We could even use an EPROM chip which could be re-written at will, giving the ultimate flexibility in function.

It is vitally important to recognize the significance of this principle as applied to digital circuitry. Whereas the half-adder built from gates or relays *processes* the input bits to arrive at a specific output, the ROM simply *remembers* what the outputs should be for any given combination of inputs.

This is not much different from the "times tables" memorized in grade school: rather than having to calculate the product of 5 times 6 (5 + 5 + 5 + 5 + 5 + 5 = 30), schoolchildren are taught to remember that 5 × 6 = 30, and then expected to recall this product from memory as needed. Likewise, rather than the logic function depending on the functional arrangement of hard-wired gates or relays (hardware), it depends solely on the data written into the memory (software).

Such a simple application, with definite outputs for every input, is called a *look-up table*, because the memory device simply "looks up" what the output(s) should to be for any given combination of inputs states.

This application of a memory device to perform logical functions is significant for several reasons:

- Software is much easier to change than hardware.
- Software can be archived on various kinds of memory media (disk, tape), thus providing an easy way to document and manipulate the function in a "virtual" form; hardware can only be "archived" abstractly in the form of some kind of graphical drawing.
- Software can be copied from one memory device (such as the EPROM chip) to another, allowing the ability for one device to "learn" its function from another device.
- Software such as the logic function example can be designed to perform functions that would be extremely difficult to emulate with discrete logic gates (or relays!).

The usefulness of a look-up table becomes more and more evident with increasing complexity of function. Suppose we wanted to build a 4-bit adder circuit using a ROM. We'd require a ROM with 8 address lines (two 4-bit numbers to be added together), plus 4 data lines (for the signed output):

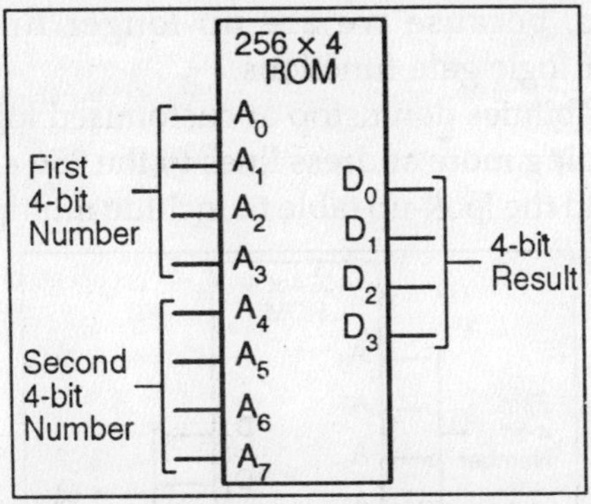

With 256 addressable memory locations in this ROM chip, we would have a fair amount of programming to do, telling it what binary output to generate for each and every combination of binary inputs. We would also run the risk of making a mistake in our programming and have it output an incorrect sum, if we weren't careful. However, the flexibility of being able to configure this function (or any function) through software alone generally outweighs that costs.

Consider some of the advanced functions we could implement with the above "adder." We know that when we add two sets of numbers in 2's complement signed notation, we risk having the answer overflow. For instance, if we try to add 0111 (decimal 7) to 0110 (decimal 6) with only a 4-bit number field, the answer we'll get is 1001 (decimal -7) instead of the correct value, 13 (7 + 6), which cannot be expressed using 4 signed bits. If we wanted to, we could avoid the strange answers given in overflow conditions by programming this look-up table circuit to output something else in conditions where we know overflow will occur (that is, in any case where the real sum would exceed +7 or -8). One alternative might be to programme the ROM to output the quantity 0111 (the maximum positive value that can be represented with 4 signed bits), or any other value that we determined to be more appropriate for the application than the typical overflowed "error" value that a regular adder circuit would output. It's all up to the programmer to decide what he or she wants this circuit to do, because we are no longer limited by the constraints of logic gate functions.

The possibilities don't stop at customised logic functions, either. By adding more address lines to the 256 × 4 ROM chip, we can expand the look-up table to include multiple functions:

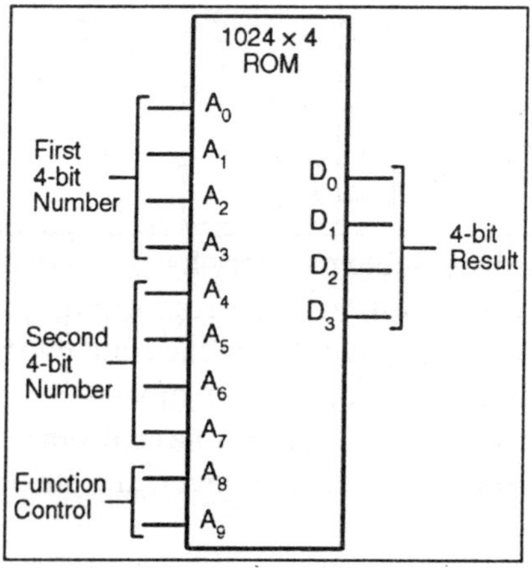

With two more address lines, the ROM chip will have 4 times as many addresses as before (1024 instead of 256). This ROM could be programmed so that when A8 and A9 were both low, the output data represented the *sum* of the two 4-bit binary numbers input on address lines A0 through A7, just as we had with the previous 256 × 4 ROM circuit.

For the addresses A8=1 and A9=0, it could be programmed to output the *difference* (subtraction) between the first 4-bit binary number (A0 through A3) and the second binary number (A4 through A7).

For the addresses A8=0 and A9=1, we could programme the ROM to output the difference (subtraction) of the two numbers in reverse order (second - first rather than first - second), and finally, for the addresses A8=1 and A9=1, the ROM could be programmed to compare the two inputs and output an indication of equality or inequality.

What we will have then is a device that can perform four different arithmetical operations on 4-bit binary numbers, all by "looking up" the answers programmed into it.

If we had used a ROM chip with more than two additional address lines, we could programme it with a wider variety of functions to perform on the two 4-bit inputs.

There are a number of operations peculiar to binary data (such as parity check or Exclusive-ORing of bits) that we might find useful to have programmed in such a look-up table.

Devices such as this, which can perform a variety of arithmetical tasks as dictated by a binary input code, are known as*Arithmetic Logic Units* (ALUs), and they comprise one of the essential components of computer technology.

Although modern ALUs are more often constructed from very complex combinational logic (gate) circuits for reasons of speed, it should be comforting to know that the exact same functionality may be duplicated with a "dumb" ROM chip programmed with the appropriate look-up table(s).

In fact, this exact approach was used by IBM engineers in 1959 with the development of the IBM 1401 and 1620 computers, which used look-up tables to perform addition, rather than binary adder circuitry.

The machine was fondly known as the "CADET," which stood for "Can't Add, Doesn't Even Try."

A very common application for look-up table ROMs is in control systems where a custom mathematical function needs to be represented.

Such an application is found in computer-controlled fuel injection systems for automobile engines, where the proper air/fuel mixture ratio for efficient and clean operation changes with several environmental and operational variables.

Tests performed on engines in research labouratories determine what these ideal ratios are for varying conditions of engine load, ambient air temperature, and barometric air pressure.

The variables are measured with sensor transducers, their analog outputs converted to digital signals with A/D circuitry, and those parallel digital signals used as address inputs to a high-capacity ROM chip programmed to output the optimum digital value for air/fuel ratio for any of these given conditions.

Sometimes, ROMs are used to provide one-dimensional look-up table functions, for "correcting" digitized signal values so that they more accurately represent their real-world significance.

An example of such a device is a *thermocouple transmitter*, which measures the millivoltage signal generated by a junction of dissimilar metals and outputs a signal which is supposed to *directly* correspond to that junction temperature.

Unfortunately, thermocouple junctions do not have perfectly linear temperature/voltage responses, and so the raw voltage signal is not perfectly proportional to temperature.

By digitizing the voltage signal (A/D conversion) and sending that digital value to the address of a ROM programmed with the necessary correction values, the ROM's programming could eliminate some of the nonlinearity of the thermocouple's temperature-to-millivoltage relationship, so that the final output of the device would be more accurate. The popular instrumentation term for such a look-up table is a digital *characterizer*.

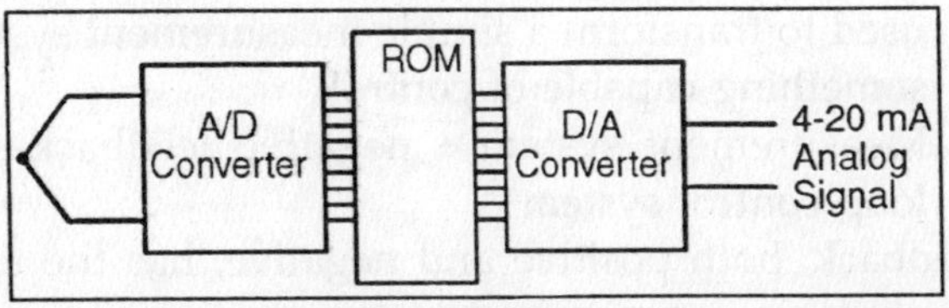

Another application for look-up tables is in special code translation. A 128 × 8 ROM, for instance, could be used to translate 7-bit ASCII code to 8-bit EBCDIC code:

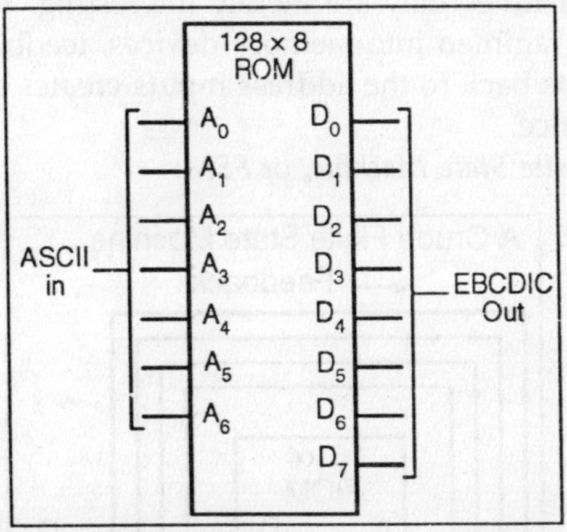

Again, all that is required is for the ROM chip to be properly programmed with the necessary data so that each valid ASCII input will produce a corresponding EBCDIC output code.

FINITE-STATE MACHINES

Feedback is a fascinating engineering principle. It can turn a rather simple device or process into something substantially more complex. We've seen the effects of feedback intentionally integrated into circuit designs with some rather astounding effects:

- Comparator + negative feedback→ controllable-gain amplifier
- Comparator + positive feedback→ comparator with hysteresis
- Combinational logic + positive feedback → multivibrator

- In the field of process instrumentation, feedback is used to transform a simple measurement system into something capable of control:
- Measurement system + negative feedback→closed-loop control system

Feedback, both positive and negative, has the tendency to add whole new dynamics to the operation of a device or system.

Sometimes, these new dynamics find useful application, while other times they are merely interesting. With look-up tables programmed into memory devices, feedback from the data outputs back to the address inputs creates a whole new type of device.

The *Finite State Machine*, or *FSM*:

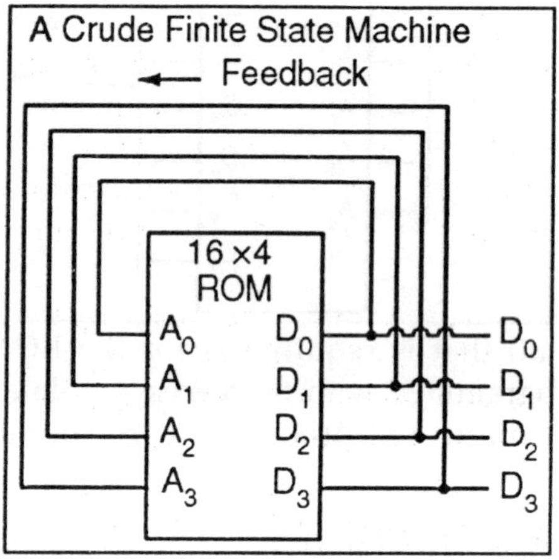

The above circuit illustrates the basic idea: the data stored at each address becomes the next storage location that the ROM gets addressed to.

The result is a specific sequence of binary numbers (following the sequence programmed into the ROM) at the output, over time.

To avoid signal timing problems, though, we need to connect the data outputs back to the address inputs through a 4-bit D-type flip-flop, so that the sequence takes place step by step to the beat of a controlled clock pulse:

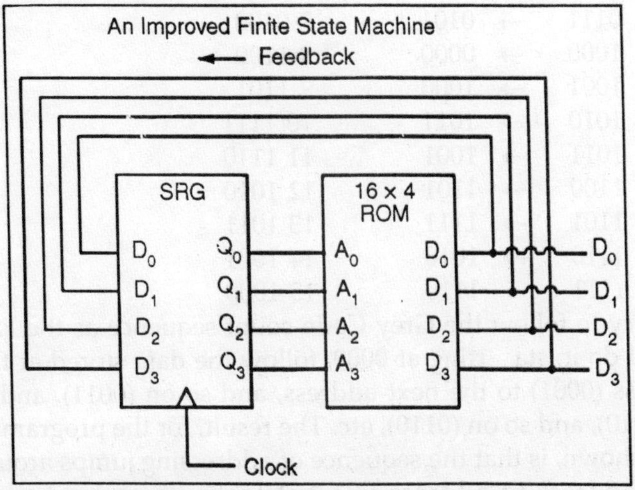

An analogy for the workings of such a device might be an array of post-office boxes, each one with an identifying number on the door (the address), and each one containing a piece of paper with the address of another P.O. box written on it (the data). A person, opening the first P.O. box, would find in it the address of the next P.O. box to open. By storing a particular pattern of addresses in the P.O. boxes, we can dictate the sequence in which each box gets opened, and therefore the sequence of which paper gets read.

Having 16 addressable memory locations in the ROM, this Finite State Machine would have 16 different stable "states" in which it could latch. In each of those states, the identity of the next state would be programmed in to the ROM, awaiting the signal of the next clock pulse to be fed back to the ROM as an address. One useful application of such an FSM would be to generate an arbitrary count sequence, such as Grey Code:

Address	→	*Data*	*Grey Code count sequence:*
0000	→	0001	0 0000
0001	→	0011	1 0001
0010	→	0110	2 0011
0011	→	0010	3 0010
0100	→	1100	4 0110
0101	→	0100	5 0111
0110	→	0111	6 0101

0111	→	0101	7 0100
1000	→	0000	8 1100
1001	→	1000	9 1101
1010	→	1011	10 1111
1011	→	1001	11 1110
1100	→	1101	12 1010
1101	→	1111	13 1011
1110	→	1010	14 1001
1111	→	1110	15 1000

Try to follow the Grey Code count sequence as the FSM would do it: starting at 0000, follow the data stored at that address (0001) to the next address, and so on (0011), and so on (0010), and so on (0110), etc. The result, for the programme table shown, is that the sequence of addressing jumps around from address to address in what looks like a haphazard fashion, but when you check each address that is accessed, you will find that it follows the correct order for 4-bit Grey code. When the FSM arrives at its last programmed state (address 1000), the data stored there is 0000, which starts the whole sequence over again at address 0000 in step with the next clock pulse.

We could expand on the capabilities of the above circuit by using a ROM with more address lines, and adding more programming data:

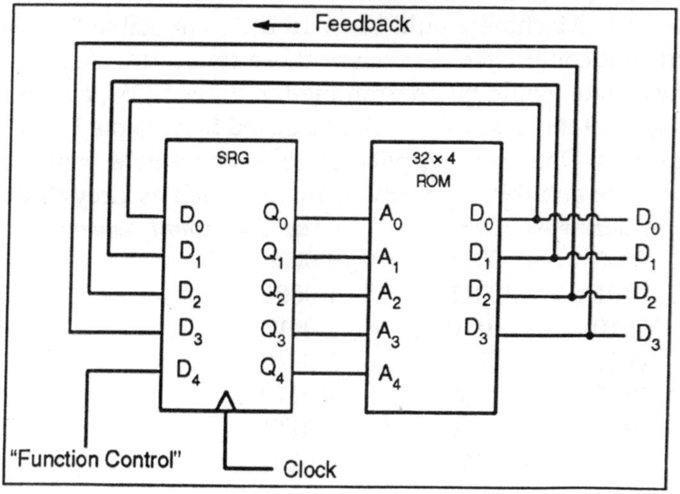

Now, just like the look-up table adder circuit that we turned into an Arithmetic Logic Unit (+, –, ×,/functions) by utilizing more address lines as "function control" inputs, this FSM counter can be used to generate more than one count sequence, a different sequence programmed for the four feedback bits (A0 through A3) for each of the two function control line input combinations (A4 = 0 or 1).

Address	→	*Data*	*Address*	→	*Data*
00000	→	0001	10000	→	0001
00001	→	0010	10001	→	0011
00010	→	0011	10010	→	0110
00011	→	0100	10011	→	0010
00100	→	0101	10100	→	1100
0010	→	0110	10101	→	0100
00110	→	0111	10110	→	0111
0011	→	1000	10111	→	0101
01000	→	1001	11000	→	0000
01001	→	1010	11001	→	1000
01010	→	1011	11010	→	1011
01011	→	1100	11011	→	1001
01100	→	1101	11100	→	1101
01101	→	1110	11101	→	1111
01110	→	1111	11110	→	1010
01111	→	0000	11111	→	1110

If A4 is 0, the FSM counts in binary; if A4 is 1, the FSM counts in Grey Code. In either case, the counting sequence is arbitrary: determined by the whim of the programmer. For that matter, the counting sequence doesn't even have to have 16 steps, as the programmer may decide to have the sequence recycle to 0000 at any one of the steps at all. It is a completely flexible counting device, the behaviour strictly determined by the software (programming) in the ROM.

We can expand on the capabilities of the FSM even more by utilizing a ROM chip with additional address input and data output lines.

Take the following circuit, for example:

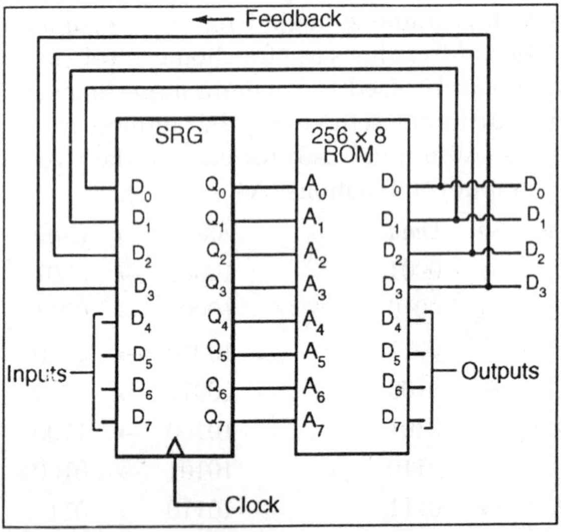

Here, the D0 through D3 data outputs are used exclusively for feedback to the A0 through A3 address lines. Date output lines D4 through D7 can be programmed to output something other than the FSM's "state" value. Being that four data output bits are being fed back to four address bits, this is still a 16-state device.

However, having the output data come from other data output lines gives the programmer more freedom to configure functions than before.

In other words, this device can do far more than just count! The programmed output of this FSM is dependent not only upon the state of the feedback address lines (A0 through A3), but also the states of the input lines (A4 through A7). The D-type flip/flop's clock signal input does not have to come from a pulse generator, either. To make things more interesting, the flip/flop could be wired up to clock on some external event, so that the FSM goes to the next state only when an input signal tells it to.

Now we have a device that better fulfills the meaning of the word "programmable." The data written to the ROM is a programme in the truest sense: the outputs follow a pre-established order based on the inputs to the device and which "step" the device is on in its sequence.

This is very close to the operating design of the *Turing Machine,* a theoretical computing device invented by Alan Turing, mathematically proven to be able to solve any known arithmetic problem, given enough memory capacity.

DECODERS

Decoders can detect a code and activate a single output to signal the presence of that code. Decoders have many applications, from producing system alerts in alarm systems to performing the task of driving multiple devices in microprocessor systems (e.g. memory).

BASIC BINARY DECODER

The function of the binary decoder is to determine if a given input combination has occurred. For example, if we wish to detect that 1011 occurs on the inputs of a digital circuit we must design a decoder which only outputs '1' for this instance. Accordingly, a 4-input AND gate and an Inverter may be employed as illustrated in.

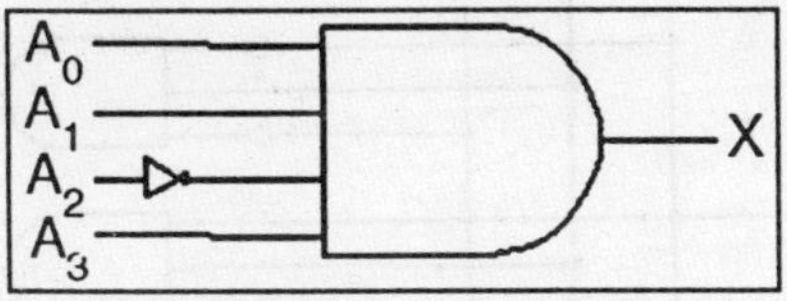

Fig. Decoding circuit for X=1011

THE THREE BIT BINARY DECODER

In order to decode all possible combinations of three bits, eight (2^3=8) decoding logic gates are required. This type of decoder is called the *3-line-to-8-line* decoder because they are 3 inputs and 8 outputs.

Let us consider the design of such a decoder and assume that we require ACTIVE HIGH outputs. That is, for a given input combination the decoder outputs '1'. To illustrate lets consider Table which list the decoding functions and truth tables for the 3-line-to-8-line decoder.

Table. Decoding Functions and Truth Tables for the 3-line-to-8-line Decoder

Decimal Digit	Binary Inputs			Logic Function	Outputs							
					D_0	D_1	D_2	D_3	D_4	D_5	D_6	D_7
0	0	0	0	$\bar{A}_2\bar{A}_1\bar{A}_0$	1	0	0	0	0	0	0	0
1	0	0	1	$\bar{A}_2\bar{A}_1A_0$	0	1	0	0	0	0	0	0
2	0	1	0	$\bar{A}_2A_1\bar{A}_0$	0	0	1	0	0	0	0	0
3	0	1	1	$\bar{A}_2A_1A_0$	0	0	0	1	0	0	0	0
4	1	0	0	$A_2\bar{A}_1\bar{A}_0$	0	0	0	0	1	0	0	0
5	1	0	1	$A_2\bar{A}_1A_0$	0	0	0	0	0	1	0	0
6	1	1	0	$A_2A_1\bar{A}_0$	0	0	0	0	0	0	1	0
7	1	1	1	$A_2A_1A_0$	0	0	0	0	0	0	0	1

Now, we can develop a decoder based on each logic function and implement the SOP logic circuit. This is illustrated below in Figure.

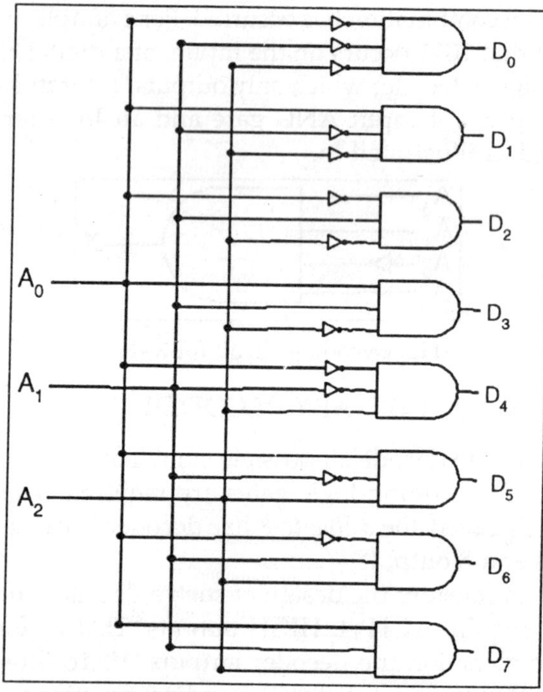

Fig. Internal Circuitry for 3-line-to-8-line decoder

It is far more convenient to use the logic symbol for the3-line-to-8-line decoder as illustrated in Figure rather than repeating the complex internal circuitry each time.

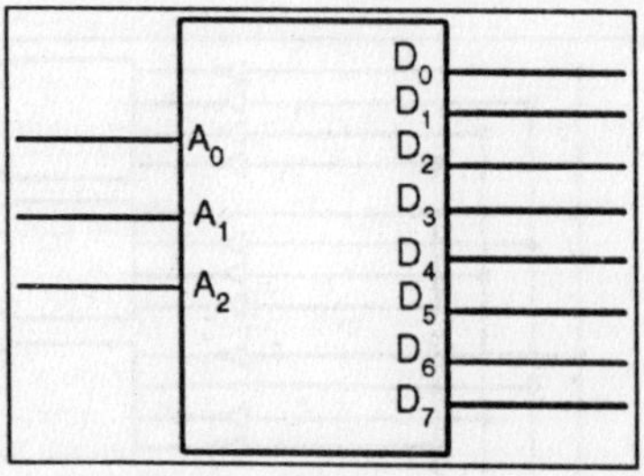

Fig. Logic Symbol for 3-line-to-8-line decoder

Note the 3-line-to-8-line is sometimes referred to as the 3 × 8 decoder

THE BCD-TO-DECIMAL DECODER

The BCD-to-decimal decoder converts each BCD code to its decimal equivalent. The technique employed is very similar to the one used in developing the 3-line-to-8-line decoder. Again assuming active high outputs are required, Table lists the decoding functions for BCD-to-decimal decoder

Table.Decoding Functions for the BCD-to-Decimal Decoder

Decimal Digit	Binary Inputs				Logic Function
0	0	0	0	0	$\bar{A}_3\bar{A}2\bar{A}_1\bar{A}_0$
1	0	0	0	1	$\bar{A}_3\bar{A}_2\bar{A}_1A_0$
2	0	0	1	0	$\bar{A}_3\bar{A}_2A_1\bar{A}_0$
3	0	0	1	1	$\bar{A}_3\bar{A}_2A_1A_0$
4	0	1	0	0	$\bar{A}_3A_2\bar{A}_1\bar{A}_0$
5	0	1	0	1	$\bar{A}_3A_2\bar{A}_1A_0$
6	0	1	1	0	$\bar{A}_3A_2A_1\bar{A}_0$
7	0	1	1	1	$\bar{A}_3A_2A_1A_0$
8	1	0	0	0	$A_3\bar{A}_2\bar{A}_1\bar{A}_0$
6	1	0	0	1	$A_3\bar{A}_2\bar{A}_1A_0$

Now, we can develop a decoder based on each logic function and implement the SOP logic circuit. This is illustrated.

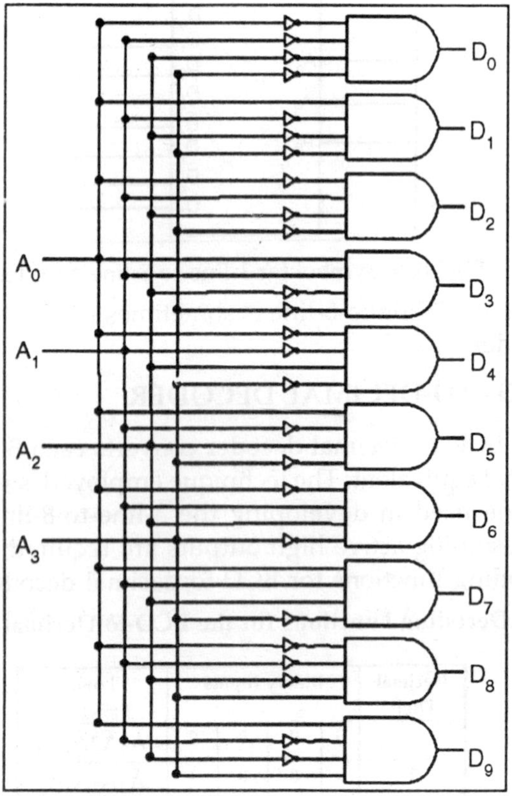

Fig. Logic Circuit for BCD Decoder

ADDER

HALF ADDER

Recall that the basic rules of binary addition are as indicated below in Table 2-9. A circuit known as the half-adder carries out these basic operations. The half-adder, illustrated in Figure, accepts two binary digits and produces the sum and carry digits.

Table.Binary Arithmetic Addition Rules

A+B	Carry	Sum
0+0	0	0
0+1	0	1
1+0	0	1
1+1	1	0

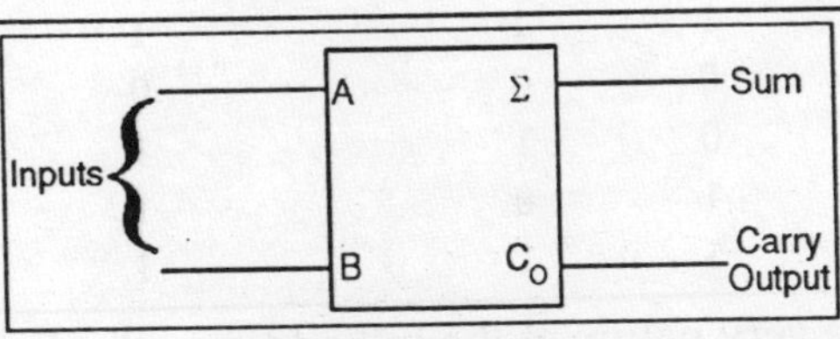

Fig.Half-Adder

Observe from Table that we may derive logical expressions for the Sum (S) and carry output (C_O) bits. Namely, $C_O = AB$, and $\Sigma = A \oplus B$. From these two expressions, the implementation of the half-adder function is developed as illustrated in Figure.

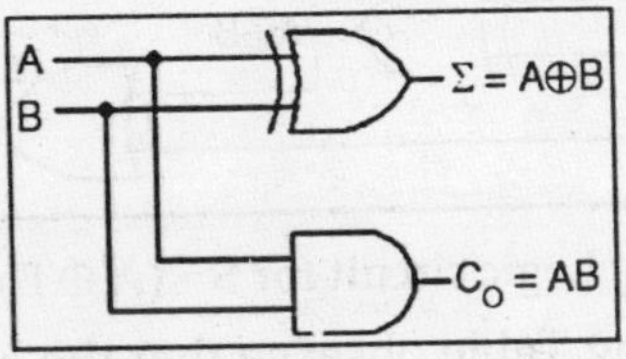

Fig.Logic Circuit for half-adder

FULL ADDER

The second basic category of adder is the full-adder. This combinational circuit performs the arithmetic addition of three input bits. The noticeable difference between the full- and the half-adder is the ability of the former to handle input carries (C_I). The logical symbol for the half-adder is shown in Figure and the associated truth table in Table.

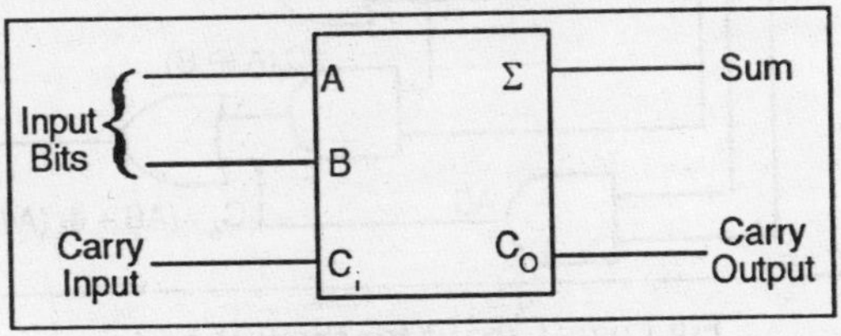

Fig.Full-Adder

Table.Full-adder Truth Table

Input			Output	
A	B	C_I	C_O	S
0	0	0	0	0
0	0	1	0	1
0	1	0	0	1
0	1	1	1	0
1	0	0	0	1
1	0	1	1	0
1	1	0	1	0
1	1	1	1	1

By the very nature of the full adder we know that the two input bits must be added to the carry input bit. Recall that for the half-adder the sum of A and B is the XOR of those two variables, $\Sigma = A \oplus B$. Similarly, for the three variables A, B and C_I the sum becomes $\Sigma = (A \oplus B) \oplus C_I$. This is illustrated in Figure.

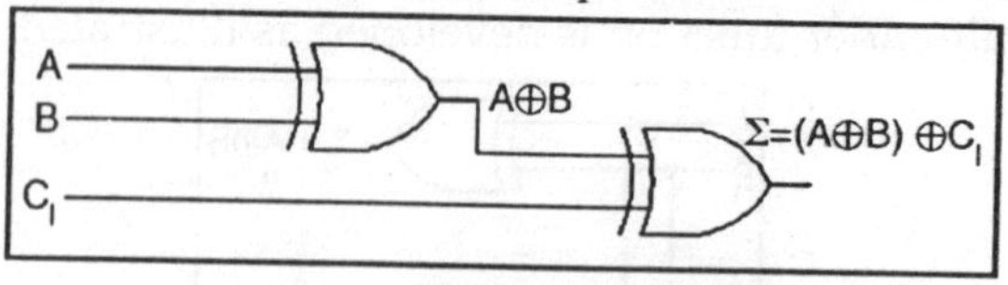

Fig. Logic circuit for $S = (A \oplus B) \oplus C_I$

By examining Table observe that the carry output (C_O) is '1' when both the inputs to the first and second XOR gates are '1'. Consequently,

$$C_O = AB + (A \oplus B)C_I .$$

Note that the complete logic circuit for the full adder is illustrated in Figure.

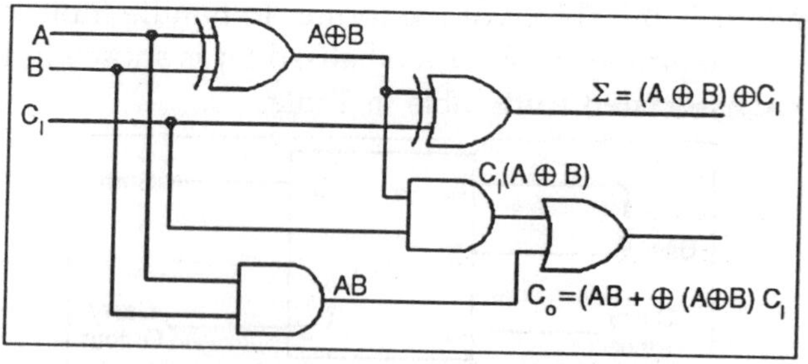

Fig. Logic Circuit for the full-adder

An alternative design of this circuit may be found by forming the SOP expression for the sum and carry functions and then implementing the results. Figure shows how two half-adders can be arranged to form a full adder.

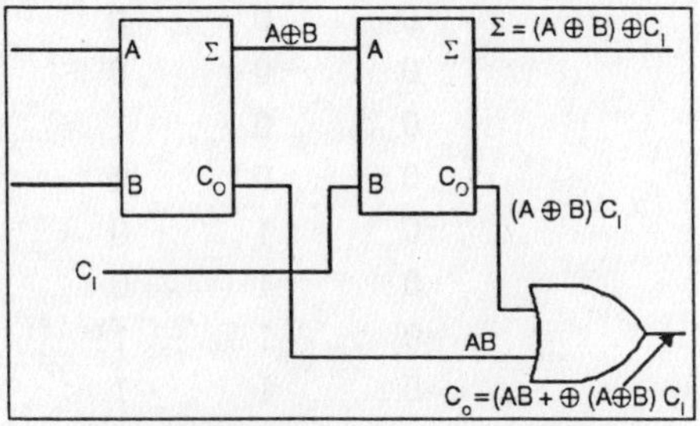

Fig.Block Diagram of full-adder

ENCODERS

An encoder is a combinational logic circuit that essentially performs the opposite function of the decoder circuit. An encoder accepts an active level (i.e. '1') on one of its inputs representing a digit (e.g. decimal or octal digit) and converts it to a coded output (e.g. binary of BCD). The process of converting from familiar symbols or numbers to a coded format is called encoding.

THE DECIMAL-TO-BCD ENCODER

Recall that with BCD the ten decimal digits (0,1,...,9) are represented by their four-digit binary counterparts. Consequently, we expect the Decimal-to-BCD encoder to have 10 inputs and 4 outputs. The logic symbol for this 10-line-to-4-line decoder is presented in Figure and the associated conversion table listed in Table with A_3representing the most significant bit.

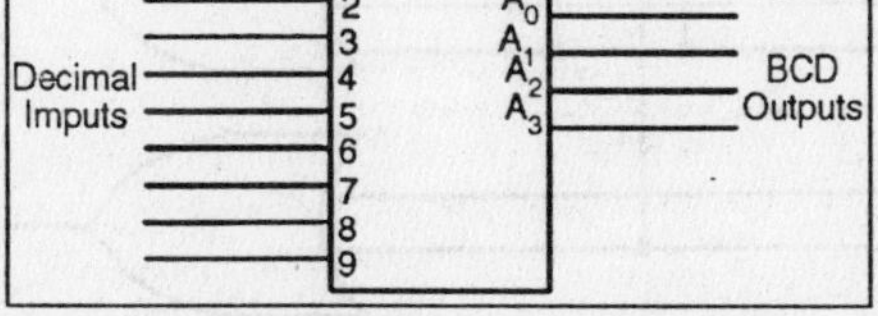

Fig.Logic Symbol for Decimal-to-BCD Encoder

Table.Decimal-to-BCD Code Table

Decimal Digit	BCD Code			
	A_3	A_2	A_1	A_0
0	0	0	0	0
1	0	0	0	1
2	0	0	1	0
3	0	0	1	1
4	0	1	0	0
5	0	1	0	1
6	0	1	1	0
7	0	1	1	1
8	1	0	0	0
9	1	0	0	1

Now, note that the MSB is defined by the function,

$$A_3 = 8 + 9$$

Similarly, A_2, A_1 and A_0 are defined by the functions,

$$A_2 = 4 + 5 + 6 + 7$$
$$A_1 = 2 + 3 + 6 + 7$$
$$A_0 = 1 + 3 + 5 + 7 + 9$$

Now we can implement the logic circuit for the decimal-to-BCD-decoder as illustrated in Figure.

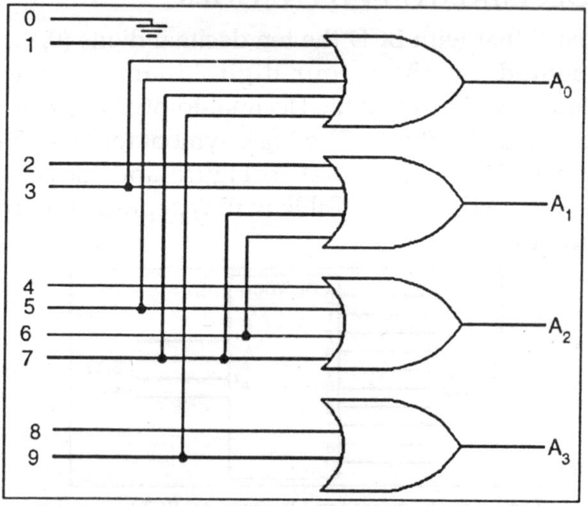

Fig. Logic Circuit for Decimal-to-BCD Encoder

Note that a connection for the decimal digit zero is not required as in this case the BCD outputs are all LOW when there are no HIGH inputs.

MULTIPLEXERS

A multiplexer (MUX) is a device that accepts data from one of many input sources for transmission over a common shared line. To achieve this the MUX has several data lines and a single output along with data-select inputs, which permit digital data on any one of the inputs to be switched to the output line. The logic symbol for a 1-of-4 data selector/ multiplexer is shown in Figure, and its associated table listed in Table.

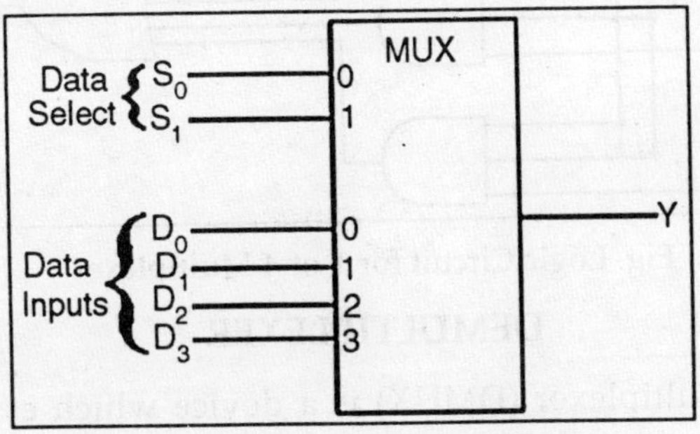

Fig.Logic Symbol for 1-of-4 Multiplexer

Table.Operation of 1-of-4 Multiplexer

Data Select Inputs		Input Selected
S_1	S_0	
0	0	D_0
0	1	D_1
1	0	D_2
1	1	D_3

Note that if a binary zero appears on the data-select lines then data on input line D_0 will appear on the output. Thus, data output Y is equal to D_0 if and only if S_1=0 and S_0=0

$$Y = D_0\overline{S}_1\overline{S}_0$$

Similarly, the data output is equal to D1, D2 and D3

for $Y = D_1\bar{S}_1S_0$, $Y = D_2S_1\bar{S}_0$ and $Y = D_3S_1S_0$, respectively. Thus the total multiplexer logic expression, formed from ORing terms is,

$$Y = D_0\bar{S}_1\bar{S}_0 + D_1\bar{S}_1S_0 + D_2S_1\bar{S}_0 + D_3S_1S_0$$

The implementation of this equation is as shown in Figure below:

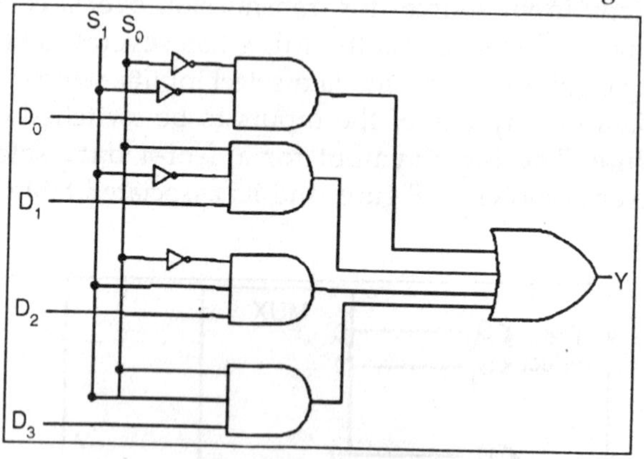

Fig. Logic Circuit for 1-of-4 Multiplexer

DEMULTIPLEXER

A demultiplexer (DMUX) is a device which essentially performs the opposite operation to the MUX. That is, it functions as an electronic switch (/data distributor) to route an incoming data signal to one of several outputs. Figure below shows the logic symbol for the 1-line-to-4-line demultiplexer circuit and Table shows list the associated Truth table. The corresponding logic circuit implementation is then shown in Figure.

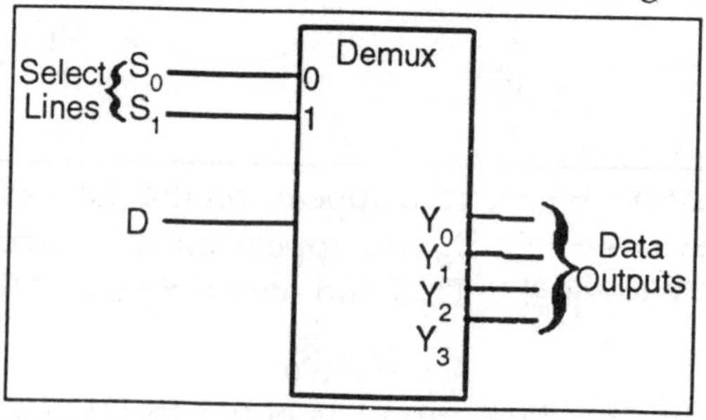

Fig. Logic Symbol for 1-line-to-4-line Demultiplexer

Table. Demultiplexer Function

Data	Address S_1	S_0	Outputs Y_0	Y_1	Y_2	Y_3
D	0	0	D	0	0	0
D	0	1	0	D	0	0
D	1	0	0	0	D	0
D	1	1	0	0	0	D

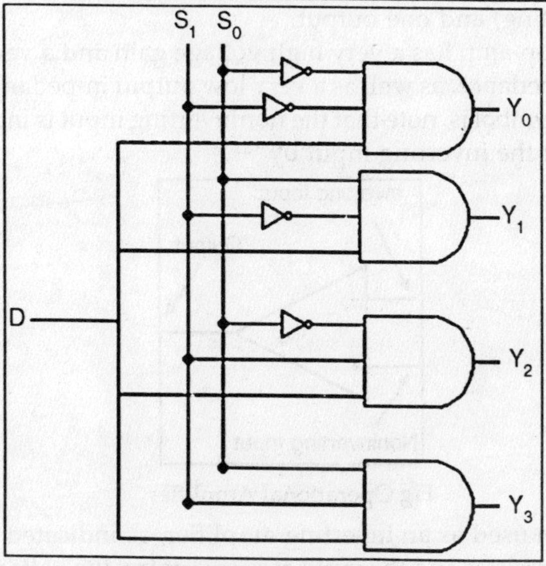

Fig.1-line-to-4-line Demultiplexer

DAC AND ADC

Digital computers and microcomputers are digital information processing systems, but information quite often is in analog form (e.g. speech, music and video signals).

To process this information with digital techniques it must first be converted from its analog to digital form. The device that does this is knows as an analog-to-digital converted (ADC or A/D converter).

In addition, since many types of electronic equipment are inherently analog devices (e.g., stereo amplifiers, radio and television receivers), there are many occasions when it is necessary to transform digital information to analog information.

This is accomplished by using a device know as a digital-to-analog converter (DAC or D/A converter).

THE OPERATIONAL AMPLIFIER

The operational amplifier, referred to as the op-amp for short, is a linear amplifier that has two inputs (inverting and noninverting) and one output.

The op-amp has a very high voltage gain and a very high input impedance, as well as a very low output impedance. The op-amp symbol is, note that the noninverting input is indicated by '-' and the inverting input by '+'.

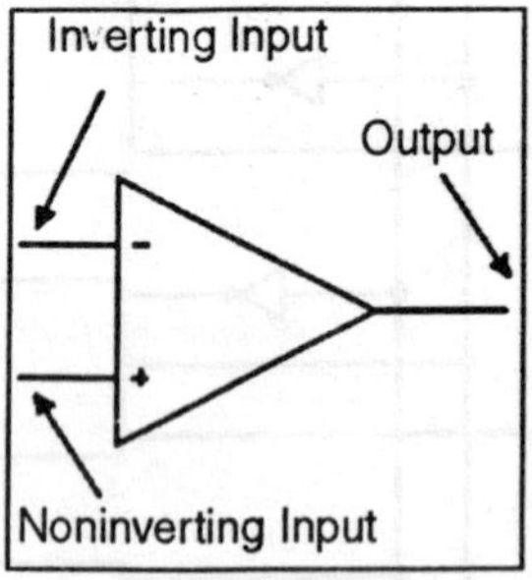

Fig.Operational Amplifier

When used as an inverting amplifier, as indicated in, the feedback resister and the input resister control the voltage gain as indicated in Equation.

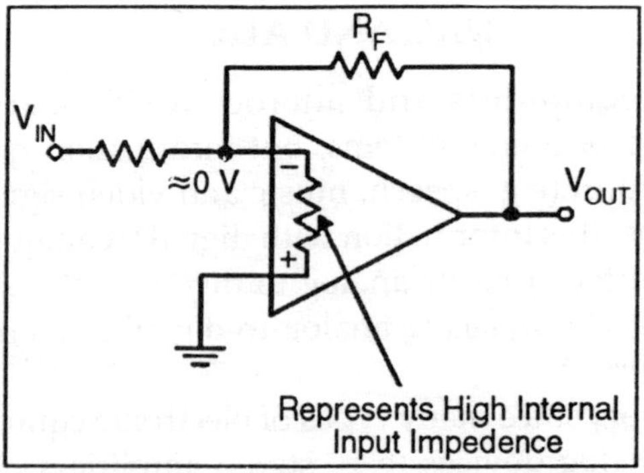

Fig. Op-amp as an Inverting Amplifier

$$\text{Gain} = \frac{V_{OUT}}{V_{IN}} = -\frac{R_F}{R_{IN}}$$

This is known as the closed-loop voltage gain because it refers to the feedback from output to input provided by R_F.

In the inverting amplifier configuration, the inverting input is approximately at ground potential (0 V). This is due to the inherent op-amp properties, the feedback mechanism and the connection of the noninverting input to the ground.

When the op-amp is used as a comparator, as, two voltages (V_{IN1} and V_{IN2}) are applied to its inputs. When these voltages differ by a very small amount, the op-amp is driven into one of its saturated states, depending on which input voltage is greater.

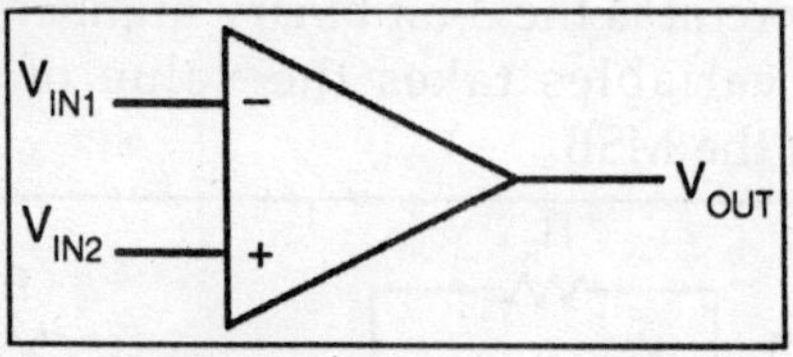

Fig.Op-amp as a Comparator

SAMPLING

Sampling is the technique of controlling the time at which information will be converted. Both digital and analog signals may be sampled. In particular, lets examine the case where analog signals are sampled. Shows the voltage follower sample-and-hold circuit.

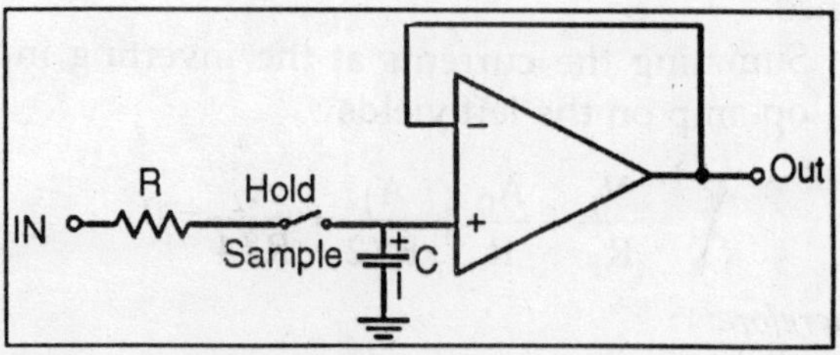

Fig. Sample-and-hold Circuit

The input signal charges the capacitor (C) and through the resistor (R) when the switch is in the "sampling" position.

At the desired instant, the switch is changed to the "hold" position isolating the input signal and leaving the input potential, at that instant, across the capacitor at the amplifier input. Ideally, this voltage would be maintained (*held*) indefinitely at the noninverting (+) input and consequently at the output.

However, finite currents at the amplifier inputs or through the sampling switch cause the voltage across C to change with time. An ideal sample-and-hold circuit would have infinitesimal sample and hold times.

DIGITAL-TO-ANALOG CONVERTERS

A simple 3-bit D/A converter is illustrated in Figure. The input to the circuit is the 3-bit binary number $A_2A_1A_0$, where each of the variables takes the value of 0V or 1V and A_2 represents the MSB.

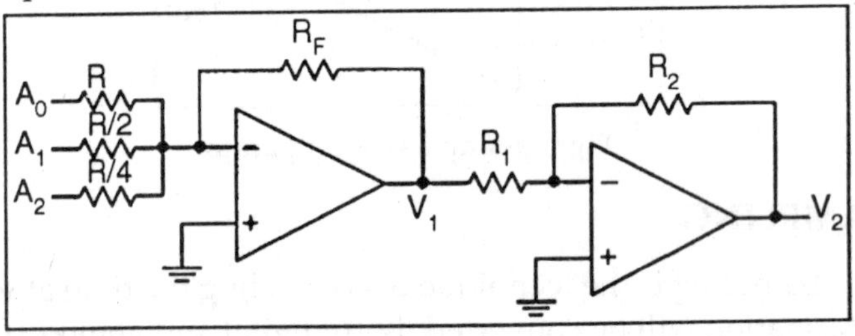

Fig. Three-bit DAC Circuit

Let us take a look at the operation of this circuit:

Note that V=IR, where V = voltage, I = current and R = resistance.

- Summing the currents at the inverting input of the op-amp on the left yields

$$\frac{V_1}{R_F}+\frac{A_0}{R}+\frac{A_1}{R/2}+\frac{A_2}{R/4}=0$$

Therefore:

$$V_1=-\frac{R_F}{R}(4A_2+2A_1+A_0)$$

- At the second stage of the op-amp the voltage gain

is given by $v_2/v_1 = -R_2/R_1$ and by the substituting for v_1 from above we get:

$$V_2 = \frac{R_2R_F}{R_1R}(4A_2 + 2A_1 + A_0)$$

- Lets consider the case where $R_1=R_F$ and $R_2=R$ the

$$V_2 = 4A_2 + 2A_1 + A_0$$

Under the circumstances when $A_2A_1A_0$=000, v_2 = 0. Similarly when $A_2A_1A_0$=111, v_2 = 7.

Performance Characteristics for D/A Converters

The performance characteristics for D/A converters include resolution, accuracy and linearity just to name a few:

- The resolution is the reciprocal of the number of discrete steps in the D/A output.
- The accuracy is a comparison of the actual output of a D/A converter with the expected output.
- Linearity is a deviation from the ideal straight-line output of a D/A converter

DESCRIPTION ABOUT AD CONVERTERS

There are a number of approaches that can be taken for the construction of an analog-to-digital converter. One technique incorporates a DAC as shown in Figure. For this device, an analog input voltage is applied to one if the inputs of a voltage comparator.

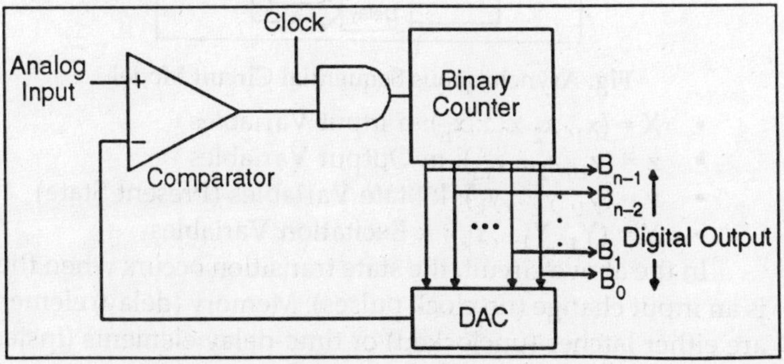

Fig. Counting ADC

Chapter 7

Concept of Asynchronous Sequential Logic

ASYNCHRONOUS CIRCUIT

An asynchronous sequential circuit is a sequential circuit whose behaviour depends only on the order in which its input signals change and can be affected at any instant of time. In addition to the FFs, a register may have combinational gates that control when and how new information is transferred into the register.

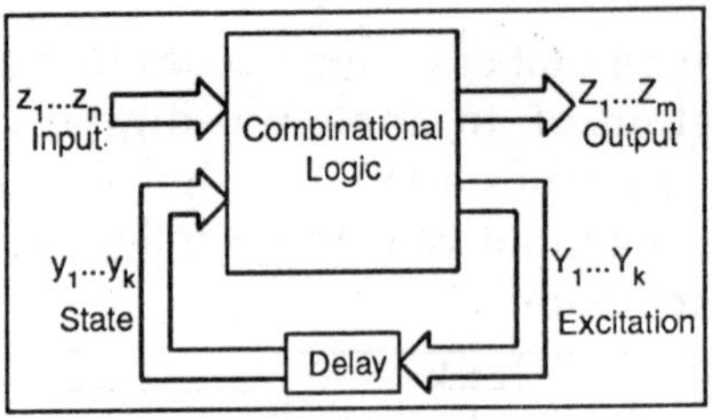

Fig. Asynchronous Sequential Circuit Model.

- $X = (x_1, x_2\ x_3..., x_n)$: n input Variables
- $z = (z_1, z_2..., z_m)$: m Output Variables
- $y = (y_1, y_2..., y_k)$: k State Variables (Present State)
- $Y = (Y_1, Y_2..., Y_k)$: k Excitation Variables

In the above circuit, the state transition occurs when there is an input change (no clock pulses). Memory (delay) elements are either latches (unclocked) or time-delay elements (instead of clocked FFs as in a synchronous sequential circuit). An asynchronous sequential circuit quite often resembles a combinational circuit with feedback. Faster and often cheaper than synchronous ones, but more difficult to design, verify, or test (due to possible timing problems involved in the feedback path).

- *Steady-state condition*: To ensure proper operation, simultaneous changes of 2 or more input variables are usually prohibited.
- *Fundamental-mode operation*: Only one input variable can change at any time, and the time between 2 input changes must be longer than the time it takes the circuit to reach a stable state.

ANALYSIS

The analysis consists of obtaining a table or a diagram that describes the sequence of internal states and outputs as a function of changes in the input variables.

- Determine all feedback loops.
- Designate each feedback-loop output with Yi and its corresponding input with yi for i = 1;:::; k, where k is the number of feedback loops.
- Derive the boolean functions for all Y's.
- Plot the transition table from the equations.

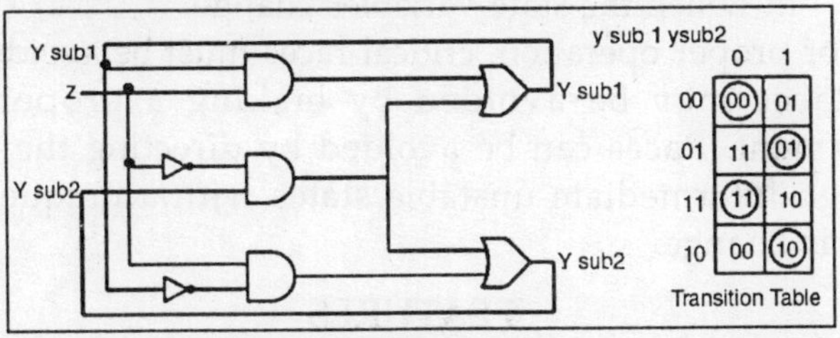

Excitation Variables: Y_1 and Y_2
Secondary Variables: y_1 and y_2

$$Y_1 = xy_1 + x'y_2$$
$$Y_1 = xy_1' + x'y_2$$

The delay associated with each feedback loop is obtained from the propagation delay between each y input and its corresponding Y output.

In an asynchronous sequential circuit, the internal state can change immediately after a change in the input.

- *Total state*: Internal state + input value.
- *Flow table*: A transition table in which states are named by letter symbols instead of specific binary values.

The flow table also includes the output values of the circuit for each

Stable State

- *Primitive flow table*: one that has only one stable state in each row. To obtain the circuit described by a flow table, it is necessary to assign to each state a distinct binary value, which converts the flow table into a transition table.
- *Race condition*: when 2 or more binary state variables change value in response to a change in an input variable.

  May cause the state variables to change in an unpredictable manner.
- *Noncritical race*: final stable state does not depend on the order in which the state variables change.
- *Critical race*: final stable state depends on the order in which the state variables change.

For proper operation, critical races must be avoided.

Races may be avoided by making a proper state assignment. Races can be avoided by directing the circuit through intermediate unstable states with a unique state-variable change.

FEATURES

- Obtain a primitive flow table.
- Reduce the flow table.
- Assign binary state variables to obtain the transition table.
- Assign output values to the dashes to obtain the output maps.
- Simplify the excitation and output functions.
- Draw the logic diagram.

We will design a gated latch circuit with 2 inputs, G (gate) and D (data), and one output, Q. The binary value at the D input is transferred to the Q output when and only when G = 1. When G falls to 0, the latch retains the value at the Q output, which does not change even if D changes.

The table of total states and the corresponding primitive flow table is shown below.

State	Input D	Input G	Output Q	Comments
a	0	1	0	D = Q
b	1	1	1	D = Q
c	0	0	0	After a or d
d	1	0	0	After c
e	1	0	1	After b or f
f	0	0	1	After e

Table of Total States

DG	00	01	11	10
a	c,–	(a)0	b,–	–,–
b	–,–	a,–	(b)1	e,–
c	(c)0	a,–	–,–	d,–
d	c,–	–,–	b,–	(d)0
e	f,–	–,–	b,–	(e)1
f	(f)1	a,–	–,–	e,–

Primitive Flow Table — States to be Merged

The next step is to reduce the primitive flow table, which is shown below.

Then,

DG	00	01	11	10
a,b,c	(c)0	(a)0	b,–	(a)0
b,e,f	(f)1	a,–	(b)1	(e)1

→

DG	00	01	11	10
a	(a)0	(a)0	b,–	(a)0
b	(b)0	a,–	(b)0	(b)0

following state assignment, we convert the flow table into a transition table (by assigning a = 0 and b = 1). From the table, the simplified O/P function (Q) and the excitation function (Y) with respect to D, G, and y can be derived using the K-map method.

Note that the don't-cares are assigned such that Q = Y

$$Y = DG + G'y; \; Q = Y$$

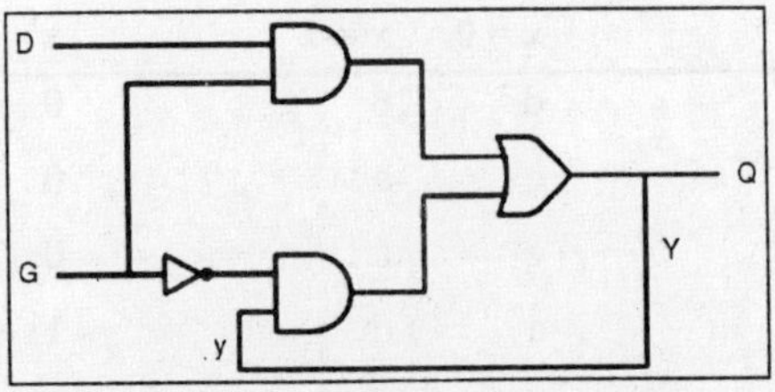

REDUCTION OF STATE

The procedure for reducing the number of internal states

in an asynchronous sequential circuit resembles the procedure that is used for synchronous circuits.

IMPLICATION TABLE

The state-reduction procedure for completely specified state tables is based on the algorithm that two states in a state table can be combined into one if they can be shown to be equivalent.

There are occasions when a pair of states do not have the same next states, but, nonetheless, go to equivalent next states.

Consider the following state table:

Present State	Next State		output	
	x = 0	x = 1	x = 0	x = 1
a	c	b	0	1
b	d	a	0	1
c	a	d	1	0
d	b	d	1	0

(a, b) imply (c, d) and (c, d) imply (a, b). Both pairs of states are equivalent; i.e., a and b are equivalent as well as c and d. The checking of each pair of states for possible equivalence in a table with a large number of states can be done systematically by means of an implication table.

This a chart that consists of squares, one for every possible pair of states, that provide spaces for listing any possible implied states.

Consider the following state table:

Present State	Next State		output	
	x = 0	x = 1	x = 0	x = 1
a	d	6	0	0
b	e	a	0	0
c	g	f	0	1
d	a	d	1	0
e	a	d	1	0
f	c	b	0	0
g	a	e	1	0

b	d.e ✓					
c	X	X				
d	X	X	X			
e	X	X	X	✓		
f	c, d ×	c, d × a, b	X	X	X	
g	X	X	X	d, e ✓	d, e ✓	X
	a	b	c	d	e	f

On the left side along the vertical are listed all the states defined in the state table except the last, and across the bottom horizontally are listed all the states except the last. The states that are not equivalent are marked with a'x' in the corresponding square, whereas their equivalence is recorded with a'?'.

Some of the squares have entries of implied states that must be further investigated to determine whether they are equivalent or not.

The step-by-step procedure of filling in the squares is as follows:

- Place a cross in any square corresponding to a pair of states whose outputs are not equal for every input.
- Enter in the remaining squares the pairs of states that are implied by the pair of states representing the squares. We do that by starting from the top square in the left column and going down and then proceeding with the next column to the right.
- Make successive passes through the table to determine whether any additional squares should be marked with a'x'. A square in the table is crossed out if it contains at least one implied pair that is not equivalent.
- Finally, all the squares that have no crosses are recorded with check marks.

The equivalent states are:

(a, b), (d, e), (d, g), (e, g).

We now combine pairs of states into larger groups of equivalent states. The last three pairs can be combined into a set of three equivalent states (d, e, g) because each one of the states in the group is equivalent to the other two.

The final partition of these states consists of the equivalent states found from the implication table, together with all the remaining states in the state table that are not equivalent to any other state:

(a, b) (c) (d, e, g) (f)

The reduced state table is:

Present State	Next State		output	
	x = 0	x = 1	x = 0	x = 1
a	d	a	0	0
c	d	f	0	1
d	a	d	1	0
f	a	c	0	0

FLOW TABLE

There are occasions when the state table for a sequential circuit is incompletely specified.

Incompletely specified states can be combined to reduce the number of states in the flow table. Such states cannot be called equivalent, but, instead they are said to be compatible.

The process that must be applied in order to find a suitable group of compatibles for the purpose of merging a flow table is divided into three steps:

- Determine all compatible pairs by using the implication table.
- Find the maximal compatibles using a merger diagram.
- Find a minimal collection of compatibles that covers all the states and is closed.

We will now proceed to show and explain the three procedural steps using the following primitive flow table:

	DG 00	01	11	10
a	C, –	(a) 0	b, –	–, –
b	–, –	a, –	(b) 1	e, –
c	(c) 0	a, –	–, –	d, –
d	c, –	–, –	b, –	(d) 0
e	f, –	–, –	b, –	(e) 0
f	(f) 0	a, –	–, –	e, –

COMPATIBLE PAIRS

Two states are compatible if in every column of the corresponding rows in the flow table, they are identical or compatible states and if there is no conflict in the output values.

The compatible pairs (γ) are:

(a, b) (a, c) (a, d) (b, e) (b, f) (c, d) (e, f)

	a	b	c	d	e
b	✓				
c	✓	d, e ×			
d	✓	d, e ×	✓		
e	C, f ×	✓	d, e × c, f ×	×	
f	C, f ×	✓	×	d, e × C, f ×	✓

MAXIMAL COMPATIBLES

The maximal compatible is a group of compatibles that

contains all the possible combinations of compatible states. The maximal compatible can be obtained from a merger diagram:

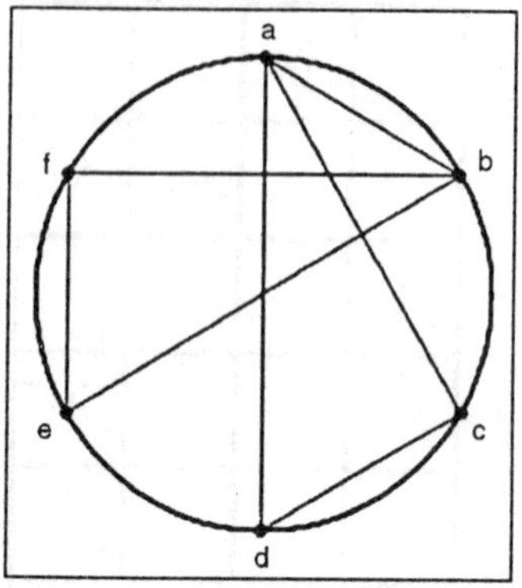

The above merger diagram is obtained from the list of compatible pairs derived from the previous implication table. A line represents a compatible pair. A triangle constitutes a compatible with three states.

The maximal compatibles are:

(a, b) (a, c, d) (b, e, f)

In the case where a state is not compatible to any other state, an isolated dot represents this state.

DIGITAL SYSTEM

While the binary numeration system is an interesting mathematical abstraction, we haven't yet seen its practical application to electronics. What makes binary numeration so important to the application of digital electronics is the ease in which bits may be represented in physical terms.

Because a binary bit can only have one of two different values, either 0 or 1, any physical medium capable of switching between two saturated states may be used to represent a bit. Consequently, any physical system capable of representing binary bits is able to represent numerical quantities, and potentially has the ability to manipulate those numbers. Electronic circuits are physical systems that lend themselves well to the representation of binary numbers. Transistors, when operated at their bias limits, may be in one of two different states: either cutoff.

If a transistor circuit is designed to maximize the probability of falling into either one of these states (and not operating in the linear, or active, mode), it can serve as a physical representation of a binary bit.

A voltage signal measured at the output of such a circuit may also serve as a representation of a single bit, a low voltage representing a binary "0" and a (relatively) high voltage representing a binary "1." Consider the following transistor circuit:

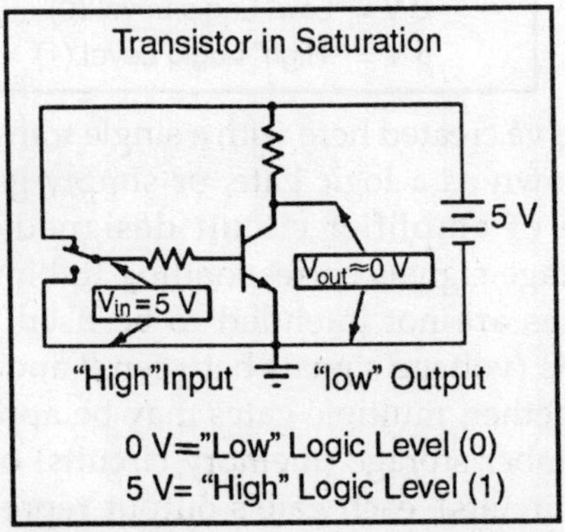

In this circuit, the transistor is in a state of saturation by virtue of the applied input voltage (5 volts) through the two-position switch. Because it's saturated, the transistor drops very little voltage between collector and emitter, resulting in an output voltage of (practically) 0 volts.

If we were using this circuit to represent binary bits, we would say that the input signal is a binary "1" and that the output signal is a binary "0."

Any voltage close to full supply voltage (measured in reference to ground, of course) is considered a "1" and a lack of voltage is considered a "0." Alternative terms for these voltage levels are high (same as a binary "1") and low (same as a binary "0").

A general term for the representation of a binary bit by a circuit voltage is logic level. Moving the switch to the other position, we apply a binary "0" to the input and receive a binary "1" at the output:

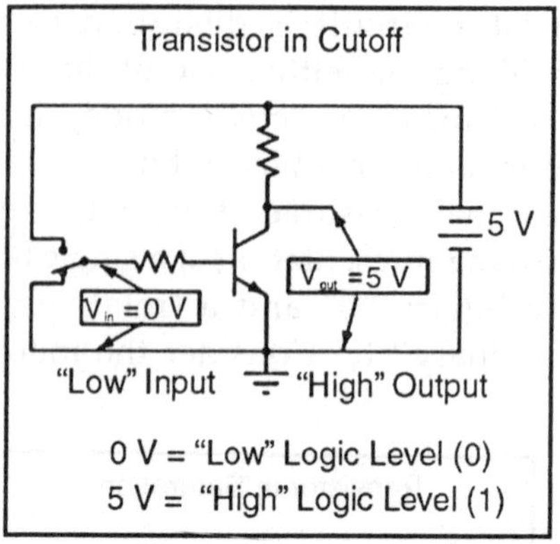

What we've created here with a single transistor is a circuit generally known as a logic gate, or simply gate. A gate is a special type of amplifier circuit designed to accept and generate voltage signals corresponding to binary 1's and 0's. As such, gates are not intended to be used for amplifying analog signals (voltage signals between 0 and full voltage).

Used together, multiple gates may be applied to the task of binary number storage (memory circuits) or manipulation (computing circuits), each gate's output representing one bit of a multi-bit binary number.

The gate shown here with the single transistor is known as an inverter, or NOT gate, because it outputs the exact opposite digital signal as what is input. For convenience, gate circuits are generally represented by their own symbols rather than by their constituent transistors and resistors.

The following is the symbol for an inverter:

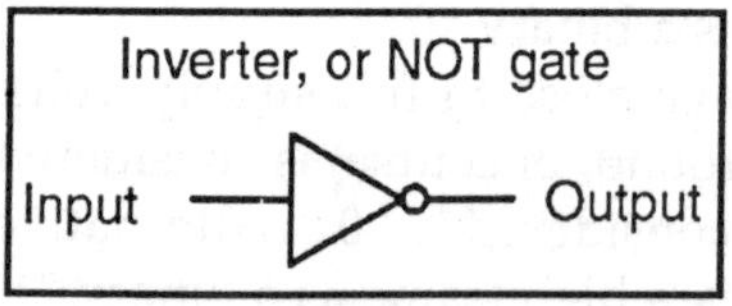

An alternative symbol for an inverter is shown here:

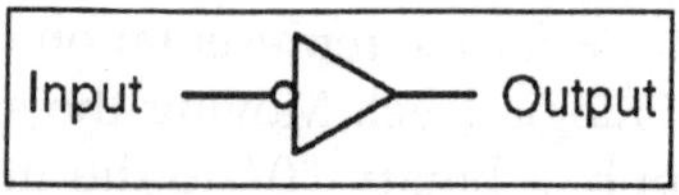

Notice the triangular shape of the gate symbol, much like

that of an operational amplifier. As was stated before, gate circuits actually are amplifiers. The small circle, or "bubble" shown on either the input or output terminal is standard for representing the inversion function.

As you might suspect, if we were to remove the bubble from the gate symbol, leaving only a triangle, the resulting symbol would no longer indicate inversion, but merely direct amplification. Such a symbol and such a gate actually do exist, and it is called a buffer, the subject of the next section.

Like an operational amplifier symbol, input and output connections are shown as single wires, the implied reference point for each voltage signal being "ground." In digital gate circuits, ground is almost always the negative connection of a single voltage source (power supply).

Dual, or "split," power supplies are seldom used in gate circuitry. Because gate circuits are amplifiers, they require a source of power to operate. Like operational amplifiers, the power supply connections for digital gates are often omitted from the symbol for simplicity's sake. If we were to show all the necessary connections needed for operating this gate, the schematic would look something like this:

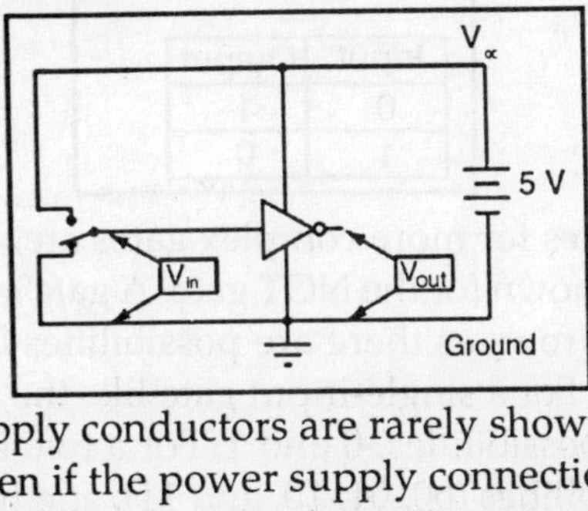

Power supply conductors are rarely shown in gate circuit schematics, even if the power supply connections at each gate are. Minimizing lines in our schematic, we get this:

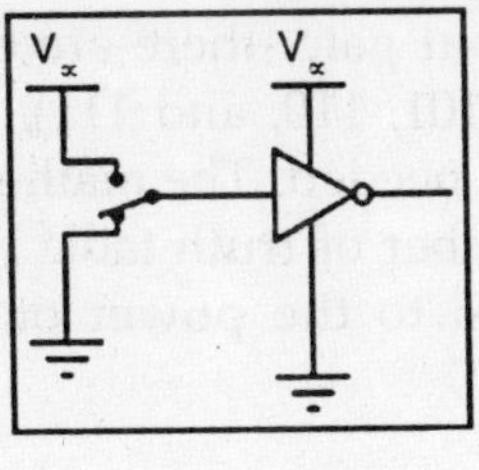

"V_{cc}" stands for the constant voltage supplied to the collector of a bipolar junction transistor circuit, in reference to ground. Those points in a gate circuit marked by the label "V_{cc}" are all connected to the same point, and that point is the positive terminal of a DC voltage source, usually 5 volts.

As we will see in other sections of this chapter, there are quite a few different types of logic gates, most of which have multiple input terminals for accepting more than one signal. The output of any gate is dependent on the state of its input(s) and its logical function.

One common way to express the particular function of a gate circuit is called a truth table. Truth tables show all combinations of input conditions in terms of logic level states (either "high" or "low," "1" or "0," for each input terminal of the gate), along with the corresponding output logic level, either "high" or "low." For the inverter, or NOT, circuit just illustrated, the truth table is very simple indeed:

NOT Gate Truth Table

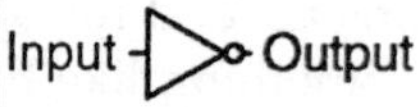

Input	Output
0	1
1	0

Truth tables for more complex gates are, of course, larger than the one shown for the NOT gate. A gate's truth table must have as many rows as there are possibilities for unique input combinations. For a single-input gate like the NOT gate, there are only two possibilities, 0 and 1. For a two input gate, there are four possibilities (00, 01, 10, and 11), and thus four rows to the corresponding truth table.

For a three-input gate, there are eight possibilities (000, 001, 010, 011, 100, 101, 110, and 111), and thus a truth table with eight rows are needed. The mathematically inclined will realise that the number of truth table rows needed for a gate is equal to 2 raised to the power of the number of input terminals.

HIGH VACUUM DIODE

A vacuum diode is an electron valve that supports current flow through it in only one direction. It is composed of two electrical "parts" or elements and the evacuated glass envelope.

Additionally, there is usually a heater, a coil of special alloy wire that can be driven to red heat by a modest application of voltage. (A vacuum diode without a heater would be a cold cathode diode.) The heater is coiled inside a cylindrical element called the cathode.

The cathode is composed of a special alloy, or, more properly is coated with an alloy, that has high thermal emission characteristics. That means that if it's heated sufficiently, electrons actually leave the metallic crystal structure of the metal (boil off) and hang around outside in the space nearby forming what is called a space charge. When we built the tube, we pumped all the air out, sealed the glass, and then fired a "getter" inside the tube to chemically bind the remaining bit of gas that was there.

The tube really is highly evacuated. That's important because we don't want all those light weight electrons bumping into gas atoms and molecules when they are hanging around in the space charge, and especially when transiting the gap between the cathode and the other element inside, the anode or plate.

If a voltage is applied across the cathode and plate, and the plate is negative, no current will flow in the tube. Electrons flow from negative to positive, but, darn it, they can't leave the plate to go to the cathode. That's called reverse biased. It's hooked up "backwards" like that. But if the cathode is connected to negative and the positive connected to the plate, electrons that were in the space charge will be driven away from the cathode and attracted to the plate. Current will flow in this direction. That's forward bias.

The hot cathode emits electrons in droves. Many more than are needed. They build up in the volume around the cathode and form a large pool of electrons just waiting for something positive to happen. This pool of electrons is called the space charge. When the plate (anode) is made positive some of the electrons are attracted to it. They impact on it, are absorbed into the metal and electrons flow out of the plate connection into the battery or what ever provided the positive voltage. The other end of the voltage source must be connected back to the cathode in some way.

If the polarity is reversed which makes the plate negative with respect to the cathode the electrons in the space charge are repelled away from the plate and no conduction takes place. The plate is cold and is made of a metal that is a very poor emitter of electrons. The voltages are not nearly as high as those used by Crooks so there is no cold emission from the plate. This makes the diode conduct current in only one direction.

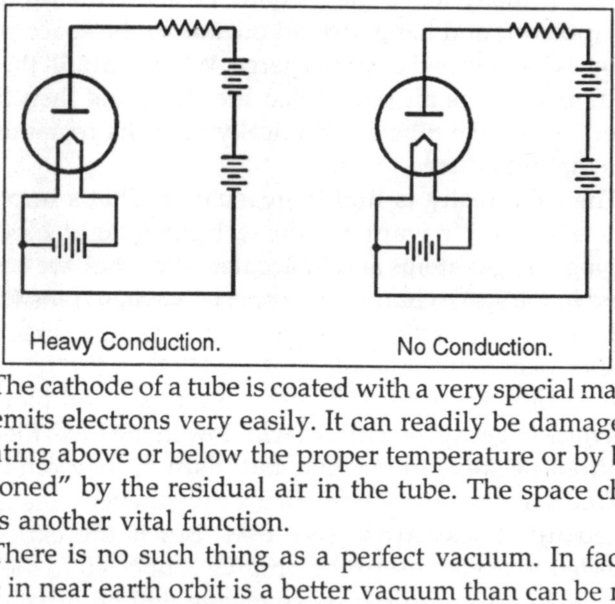

The cathode of a tube is coated with a very special material that emits electrons very easily. It can readily be damaged by operating above or below the proper temperature or by being "poisoned" by the residual air in the tube. The space charge serves another vital function.

There is no such thing as a perfect vacuum. In fact the space in near earth orbit is a better vacuum than can be made on the surface of earth. Farther out in space the vacuum is even better. What this means is there are lots of air molecules running around inside of a vacuum tube.

The space charge protects the cathode from these air molecules. Some of these molecules lose an electron and become positive ions. They are attracted to the negative cathode and if it weren't for the space charge they would impact on the surface of the cathode doing considerable damage. Over time the ability of the cathode to emit electrons would be seriously impaired rendering the tube useless. As these ions run through the space charge they gain electrons from the many collisions and become negative ions. Their direction is reversed turning them away from the delicate cathode.

INDIRECTLY HEATED CATHODES

Most vacuum tubes have a basically cylindrical configuration. Figure two shows how a filament (directly heated) cathode type diode is made along with a heater type (indirectly heated) cathode type diode.

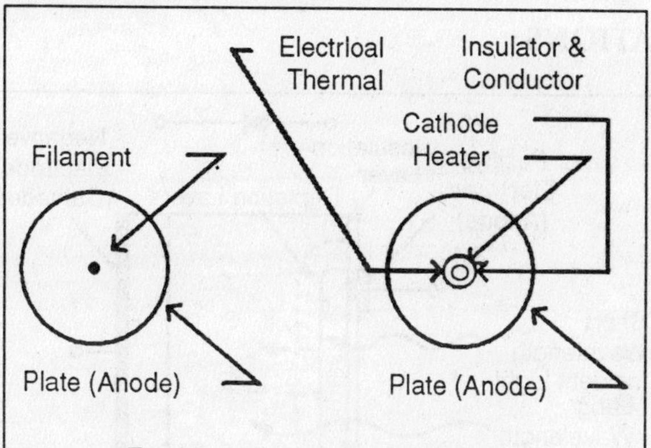

End View of Cylindrical Constructs.

PHOTODIODES

DEFINITION

Photodiodes are semiconductor light sensors that generate a current or voltage when the P-N junction in the semiconductor is illuminated by light. The term photodiode can be broadly defined to include even solar batteries, but it usually refers to sensors used to detect the intensity of light. Photodiodes can be classified by function and construction as follows:

VARIETY

- PN photodiode
- PIN photodiode
- Schottky type photodiode
- APD (Avalanche photodiode)

FEATURES

- Excellent linearity with respect to incident light
- Low noise
- Wide spectral response
- Mechanically rugged
- Compact and lightweight
- Long life

OPERATIONS

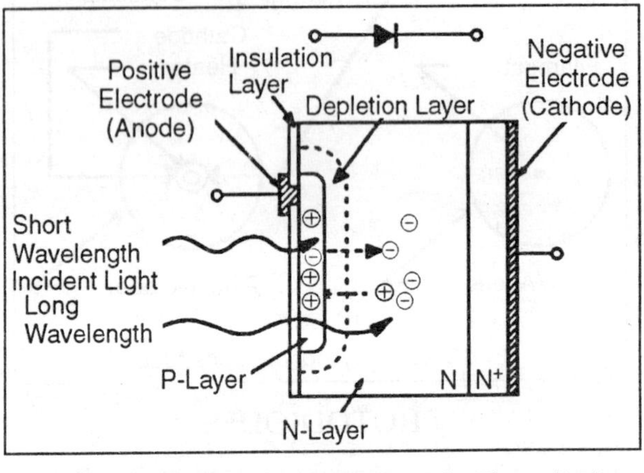

Fig. Below Shows a Cross Section of a Photodiode.

The P-layer material at the active surface and the N material at the substrate form a PN junction which operates as a photoelectric converter. The usual P-layer for a Si photodiode is formed by selective diffusion of boron, to a thickness of approximately 1 μm or less and the neutral region at the junction between the P- and N-layers is known as the depletion layer. By controlling the thickness of the outer P-layer, substrate N-layer and bottom N^+-layer as well as the doping concentration, the spectral response and frequency response can be controlled. When light strikes a photodiode, the electron within the crystal structure becomes stimulated. If the light energy is greater than the band gap energy Eg, the electrons are pulled up into the conduction band, leaving holes in their place in the valence band as shown below:

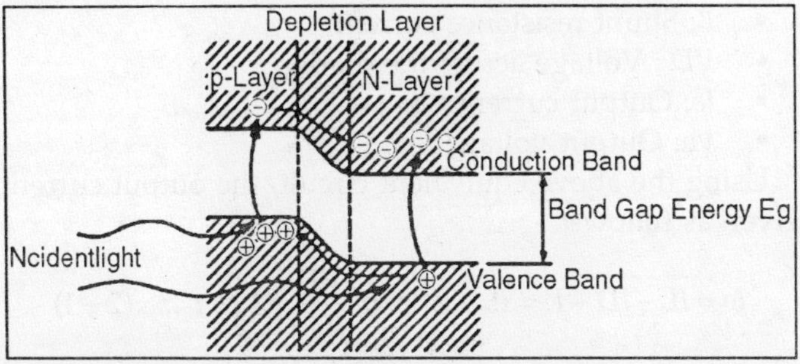

These electron-hole pairs occur throughout the P-layer, depletion layer and N-layer materials. In the depletion layer the electric field accelerates these electrons toward the N-layer and the holes toward the P-layer. Of the electron-hole pairs generated in the N-layer, the electrons, along with electrons that have arrived from the P-layer, are left in the N-layer conduction band. The holes at this time are being diffused through the N-layer up to the depletion layer while being accelerated, and collected in the P-layer valence band. In this manner, electron-hole pairs which are generated in proportion to the amount of incident light are collected in the N- and P-layers.

SILICON PHOTODIODE

An equivalent circuit of a photodiode is shown in Figure.

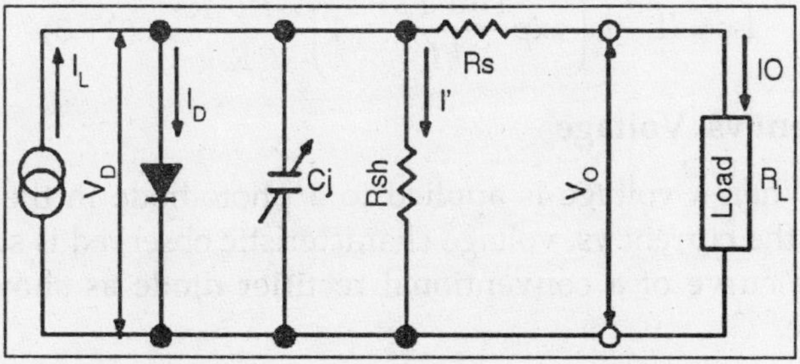

- IL : Current generated by the incident light (proportional to the amount of light)
- ID : Diode current
- Cj: Junction capacitance
- Rsh: Shunt resistance
- Rs: Series resistance
- I': Shunt resistance current
- VD : Voltage across the diode
- Io: Output current
- Vo: Output voltage

Using the above equivalent circuit, the output current Io is given as follows:

$$Io = IL - ID - I' = IL - Is\left(\exp\frac{eVD}{kT} - 1\right) - I' \text{......}(2-1)$$

- Is: Photodiode reverse saturation current
- e: Electron charge
- k: Boltzmann's constant
- T: Absolute temperature of the photodiode

The open circuit voltage Voc is the output voltage when Io equals 0. Thus Voc becomes

$$Voc = \frac{kT}{e} In\left(\frac{IL - I}{Is} + 1\right) \text{......}(2-2)$$

If I' is negligible, since Is increases exponentially with respect to ambient temperature, Voc is inversely proportional to the ambient temperature and proportional to the log of IL. However, this relationship does not hold for very low light levels.The short circuit current Isc is the output current when the load resistance RL equals 0 and Vo equals 0, yielding:

$$Isc = IL - Is\left(\exp\frac{(Isc \cdot Rs)}{kT} - 1\right) - \frac{Isc \cdot Rs}{Rsh} \text{....}(2-3)$$

Current vs. Voltage

When a voltage is applied to a photodiode in the dark state, the current vs. voltage characteristic observed is similar to the curve of a conventional rectifier diode as shown in Figure.

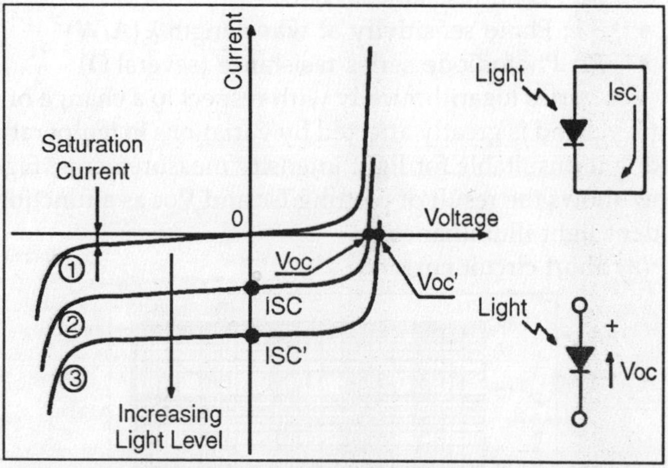

However, when light strikes the photodiode, the curve at (1) shifts to (2) and, increasing the amount of incident light shifts this characteristic curve still further to position (3) in parallel, according to the incident light intensity. As for the characteristics of (2) and (3), if the photodiode terminals are shorted, a photocurrent Isc or Isc proportional to the light intensity will flow in the direction from the anode to the cathode. If the circuit is open, an open circuit voltage Voc or Voc′ will be generated with the positive polarity at the anode.The short circuit cur.ent Isc is extremely linear with respect to the incident light level. When the incident light is within a range of 10^{-12} to 10^{-2} W, the achievable range of linearity is higher than 9 orders of magnitude, depending on the type of photodiode and its operating circuit. The lower limit of this linearity is determined by the NEP, while the upper limit depends on the load resistance and reverse bias voltage, and is given by the following equation:

$$Psat = \frac{V_{Bi} + V_R}{(Rs + R_L) \cdot S\lambda} ....(2-4)$$

- *Psat*: Input energy (W) at upper limit of linearity Psat £10 mW
- *VB*τ: Contact voltage (V) <0.2 to 0.3 V>
- *VR*: Reverse voltage (V)

- RL : Load resistance (Ω)
- $S\lambda$: Photo sensitivity at wavelength λ (A/W)
- Rs: Photodiode series resistance (several Ω)

Voc varies logarithmically with respect to a change of the light level and is greatly affected by variations in temperature, making it unsuitable for light intensity measurements. Figure below shows the result of plotting Isc and Voc as a function of incident light illuminance.

(a) Short circuit current

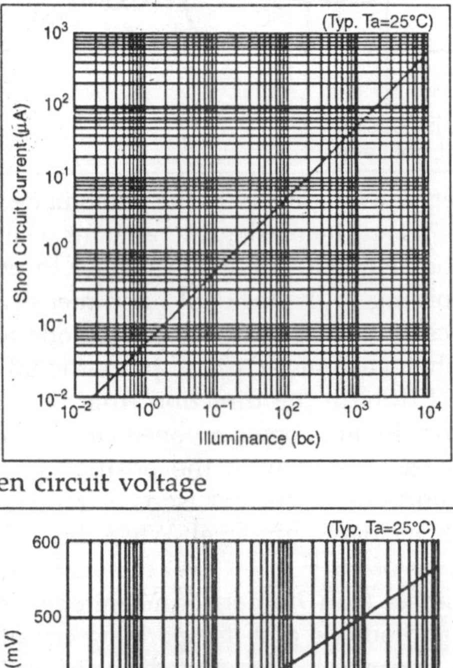

(b) Open circuit voltage

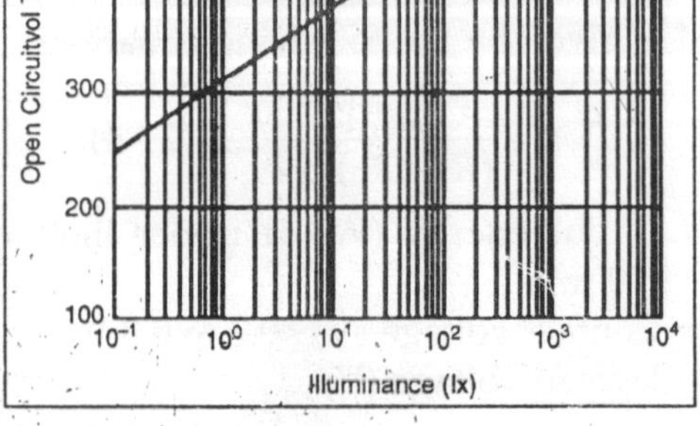

Figures (a) and (b) shown below are the methods of measuring light by measuring the photocurrent IL or Isc. In the circuit shown at (a), the voltage (Io × RL) is amplified by an amplifier with gain G, although the circuit does have limitations on its linearity according to equation (2-4). This condition is shown in Figure as shown below. Figure below (b) is a circuit using an operational amplifier. If we set the open loop gain of the operational amplifier as A, the characteristics of the feedback circuit allows the equivalent input resistance (equivalent to load resistance RL) to be,

$$\frac{Rf}{A}$$

which is several orders of magnitude smaller than Rf. Thus this circuit enables ideal Isc measurement over a wide range. For measuring a wide range, RL and Rf must be adjusted as needed.

(a) Load resistance circuit

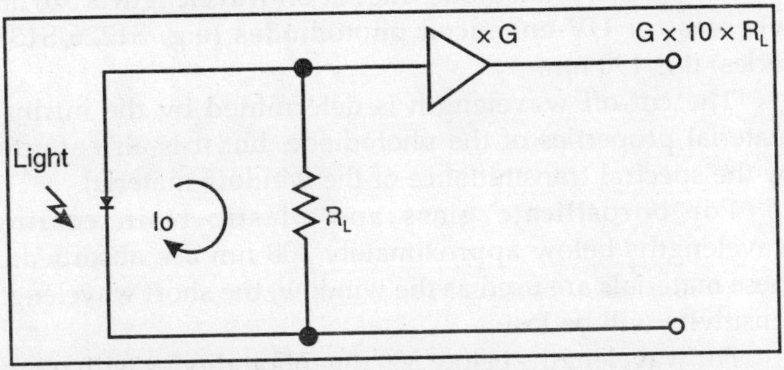

(b) Op-amp circuit

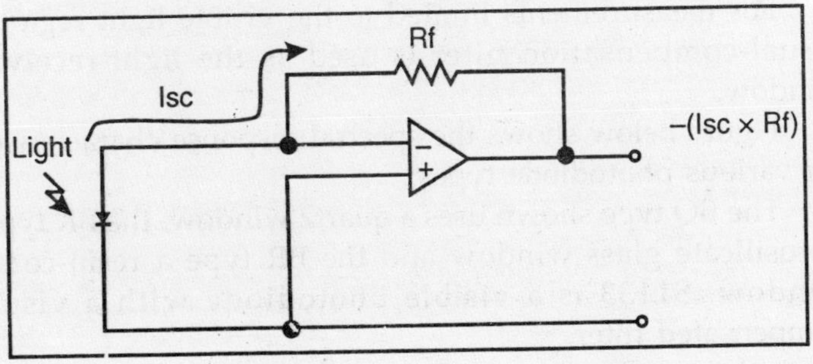

Spectral Response

As explained, when the energy of absorbed photons is lower than the band gap energy Eg, the photovoltaic effect does not occur.

The limiting wavelength lh can be expressed in terms of Eg as follows:

$$\lambda h = \frac{1240}{Eg}[nm]........(2-5)$$

At room temperatures, Eg is 1.12 eV for Si and 1.8 eV for GaAsP, so that the limiting wavelength will be 1100 nm and 700 nm, respectively. For short wavelengths, however, the degree of light absorption within the surface diffusion layer becomes very large.

Therefore, the thinner the diffusion layer is and the closer the P-N junction is to the surface, the higher the sensitivity will be.

For normal photodiodes the cut-off wavelength is 320 nm, whereas for UV-enhanced photodiodes (e.g. S1226/S1336 series) it is 190 nm.

The cut-off wavelength is determined by the intrinsic material properties of the photodiode, but it is also affected by the spectral transmittance of the window material.

For borosilicate glass and plastic resin coating, wavelengths below approximately 300 nm are absorbed. If these materials are used as the window, the short wavelength sensitivity will be lost.

For wavelengths below 300 nm, photodiodes with quartz windows are used.

For measurements limited to the visible light region, a visual-compensation filter is used as the light-receiving window.

Figure below shows the spectral response characteristics for various photodiode types.

The BQ type shown uses a quartz window, the BK type a borosilicate glass window and the BR type a resin-coated window. S1133 is a visible photodiode with a visual-compensated filter.

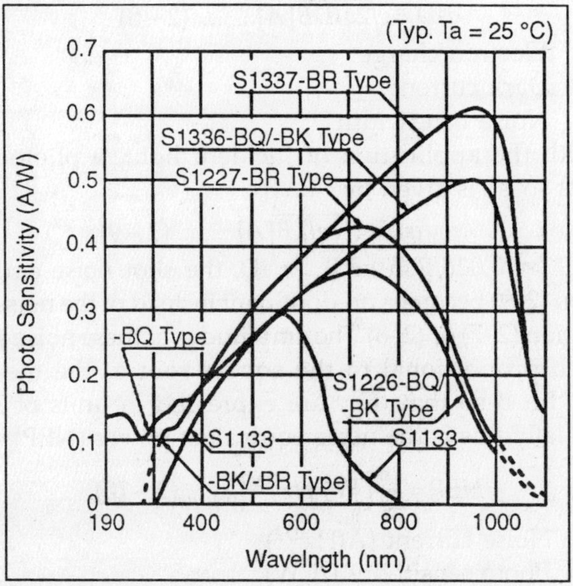

Noise Characteristic

Like other types of light sensors, the lower limits of light detection for photodiodes are determined by the noise characteristics of the device. The photodiode noise in is the sum of the thermal noise (or Johnson noise) *ij* of a resistor which approximates the shunt resistance and the shot noise isD and isL resulting from the dark current and the photocurrent.

$$in = \sqrt{ij^2 + isD^2 + isL^2}\,[A] \text{..........}(2-6)$$

ij is viewed as the thermal noise of Rsh and is given as follows:

$$ij = \sqrt{\frac{4kTB}{Rsh}}\,[A] \text{..........}(2-7)$$

K: Boltzmann's Constant

T: Absolute temperature of the element

B: Noise bandwidth

The shot noise *isD* originating from the dark current is given by,

$$isD = \sqrt{2qIDB}[A] \text{..........}(2-8)$$

q: Electron charge

ID: Dark current

B: Noise bandwidth

With the application of incident light, a photocurrent IL exists so *isL* is given by

$$isL = \sqrt{2qILB}[A] \text{..........}(2-9)$$

If IL >> 0.026/Rsh or IL >> ID, the shot noise current of equation (2–9) becomes predominant instead of the noise factor of eqaution (2–7) or (2–8).The amplitudes of these noise sources are each proportional to the square root of the measured bandwidth B so that they are expressed in units of $A/Hz^{1/2}$.Essentially this is the noise equivalent power (NEP).

$$NEP = \frac{in}{S}\left[W/Hz^{1/2}\right] \text{.........}(2-10)$$

in: Noise current ($A/Hz^{1/2}$)

S: Photo sensitivity (A/W)

Figure below shows the relationship between NEP and shunt resistance, from which a photodiode is agreement with the theoretical relationship.

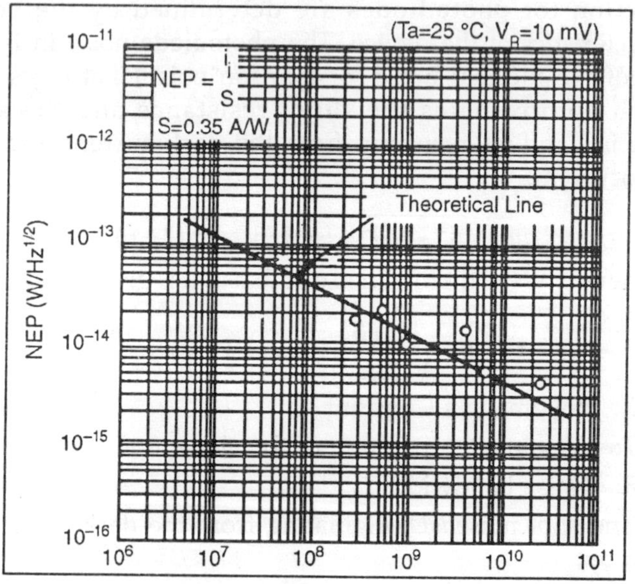

Temperature Characteristics

Ambient temperature variations greatly affect photodiode sensitivity and dark current. The cause of this is variation in the light absorption coefficient which is temperature related. For long wavelengths, sensitivity increases with increasing temperature and this increase become prominent at wavelengths longer than the peak wavelength. For short wavelengths, it decreases. Since ultraviolet enhanced photodiodes are designed to have low absorption in the short wavelength region, the temperature coefficient is extremely small at wavelengths shorter than the peak wavelength.

Figure below shows examples of temperature coefficients of photodiodes sensitivity for a variety of photodiodes types.

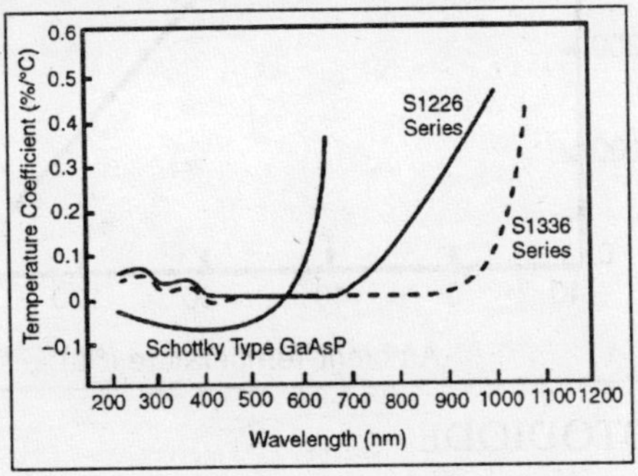

The variation occurs as a result of increasing temperatures causing electrons in the valence band to become excited, pulling them into the conduction band.

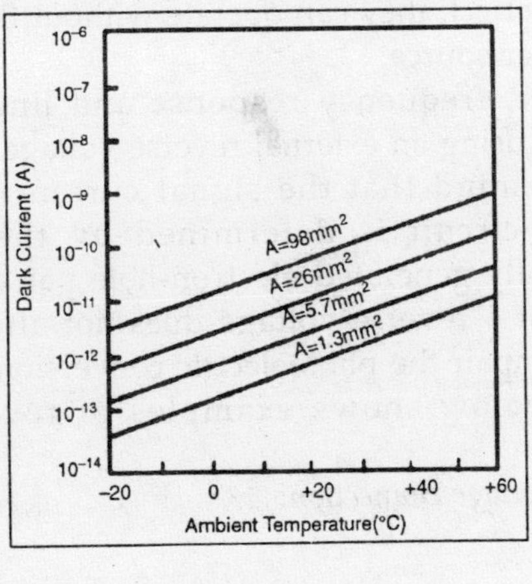

Figure below shows an example of the temperature characteristics of open-circuit voltage Vop, indicating linearity with respect to temperature change.

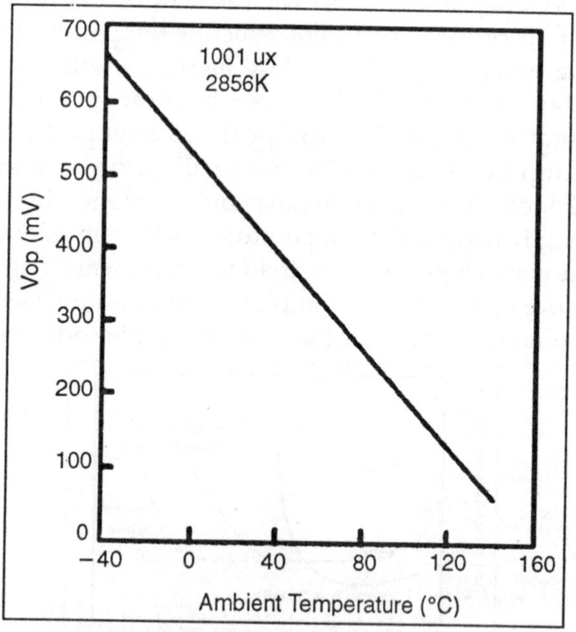

SI PIN PHOTODIODE

Reverse Voltage

Because photodiodes generate a power due to the photovoltaic effect, they can operate without the need for an external power source.

However, frequency response and linearity can be improved by using an external reverse voltage VR. It should be borne in mind that the signal current flowing in a photodiode circuit is determined by the number of photovoltaically generated electron-hole pairs and that the application of a reverse voltage does not affect the signal current nor impair the photoelectric conversion linearity.

Figure below shows examples of reverse voltage connection.

Reverse voltage connection:

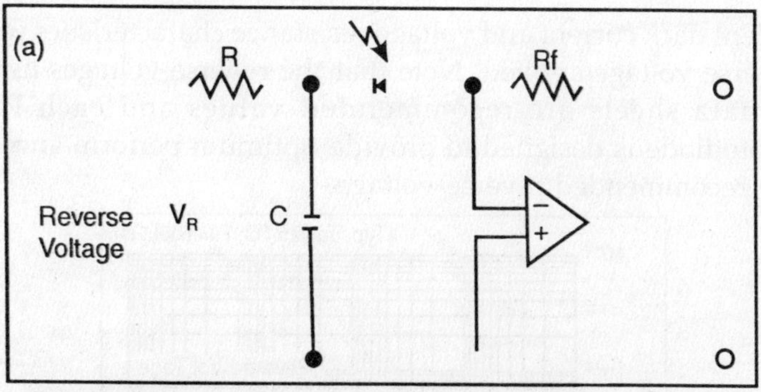

Figures below shows the effect of reverse voltage on cut-off frequency and linearity limits, respectively. While application of a reverse voltage to a photodiode is very useful in improving frequency response and linearity, it has the accompanying disadvantage of increasing dark current and noise levels along with the danger of damaging the device by excessive applied reverse voltage.

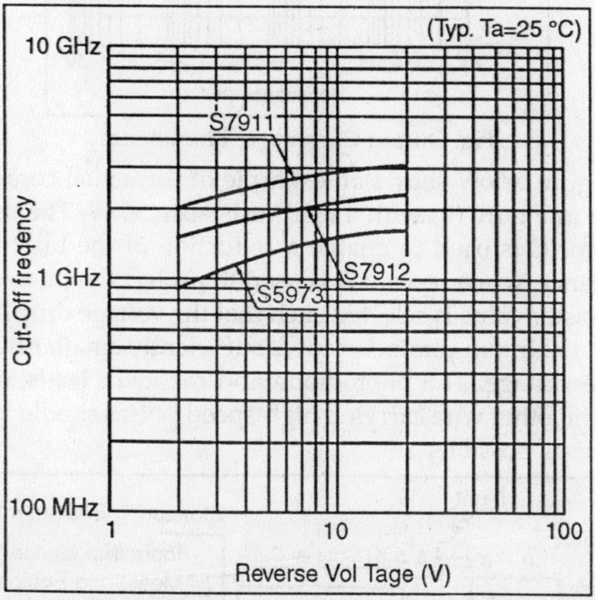

For use in applications such as optical communications and remote control which require high response speed, the PIN photodiode provides not only good response speed but excellent dark current and voltage resistance characteristics with reverse voltage applied. Note that the reverse voltages listed in data sheets are recommended values and each PIN photodiode is designed to provide optimum perform-ance at the recommended reverse voltage.

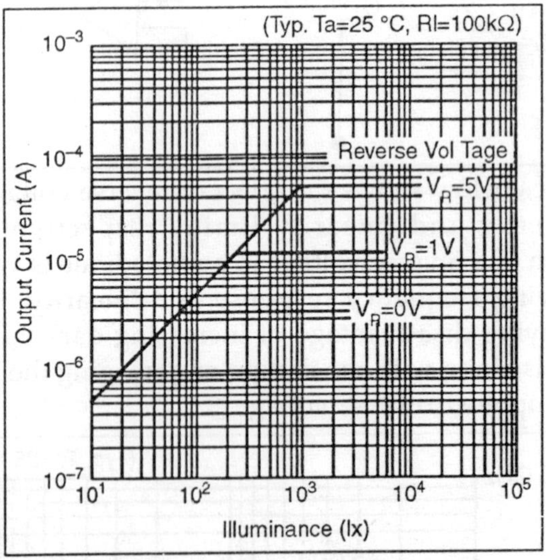

Fig. Output Current vs. illuminance

Figure below shows an example of the actual connection shown in Figure (b) with a load resistance 50 W. The ceramic capacitor C is used to enable a reduction of the bias supply impedance, while resistor R is used to protect the photodiode. The resistor value is selected such that the voltage drop caused by the maximum photocurrent is sufficiently smaller than the reverse voltage. The photodiode and capacitor leads, coaxial cable and other wire carrying high-speed pulses should be kept as short as possible.

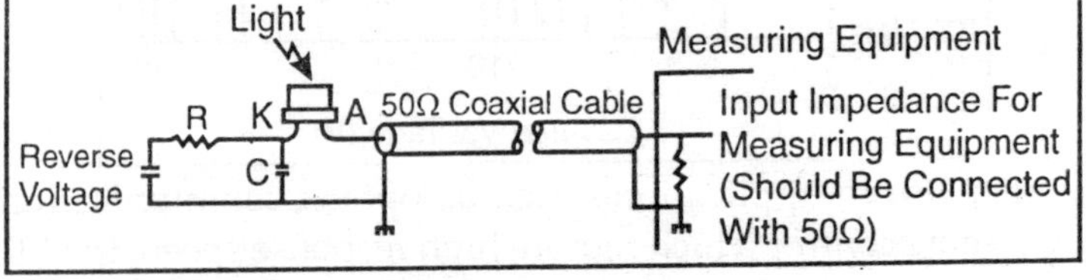

Fig.. Connection to Coaxial Cable

Response Speed and Frequency Response

The response speed of a photodiode is a measure of the time required for the accumulated charge to become an external current and is generally expressed as the rise time or cut-off frequency.

The rise time is the time required for the output signal to change from 10 % to 90 % of the peak output value and is determined by the following factors:

Terminal Capacitance Ct and Time Constant t1 of Load Resistance RL

Time constant t1 determined by the terminal capacitance Ct of the photodiode and the load resistance RL. Ct is the sum of the package capacitance and the photodiode junction capacitance. t1 is given by,

$$t1 = 2.2 \times Ct \times RL$$

To shorten t1, the design must be such that either Ct or RL is made smaller. Cj is nearly proportional to the active area A and inversely proportional to the second to third root of the depletion layer width d.

Since the depletion layer width is proportional to the product of the resistivity r of the substrate material and reverse voltage VR, the following equation is established as:

$$Cj \propto A\left\{(VR + 0.5) \times \rho\right\}^{-1/2 to -1/3}$$

Accordingly, to shorten t1, a photodiode with a small A and large r should be used with a reverse voltage applied. However, reverse voltage also increases dark current so caution is necessary for use in low-light-level detection.

Diffusion Time t2 of Carriers Generated Outside the Depletion Layer

Carriers may generate outside the depletion layer when incident light misses the P-N junction and is absorbed by the surrounding area of the photodiode chip and the substrate section which is below the depletion area. The time t2 required for these carriers to diffuse may sometimes be greater than several microseconds.

Carrier Transit Time t3 in the Depletion Layer

The transit speed vd at which the carriers travel in the depletion layer is expressed using the traveling rate μ and the electric field E developed in the depletion layer, as in vd = μE.

If we let the depletion layer width be d and the applied voltage be VR, the average electric field E=VR/d, and thus t3 can be approximated as follows:

$$t3 = d / vd = d^2 /(\mu VR)$$

To achieve a fast response time for t3, the moving distance of carriers should be short and the reverse voltage larger.

The above three factors determine the rise time tr of a photodiode and rise time tr is approximated by the following equation:

$$tr = \sqrt{t1^2 + t2^2 + t3^2}$$

PIN photodiodes and avalanche photodiodes are designed such that less carriers are generated outside the depletion layer, Ct is small and the carrier transit time in the depletion layer is short.

Therefore, these types are ideally suited for high-speed light detection.The cut-off frequency fc is the frequency at which the photodiode output decreases by 3 dB from the output at 100 kHz when the photodiode receives sinewave-modulated light from a laser diode.

The rise time tr roughly approximates this fc in the formula:

$$tr = \frac{0.35}{fc}$$

Figures (a), (b) and (c) below show examples of the response waveform and frequency response character-istics for typical photodiodes.

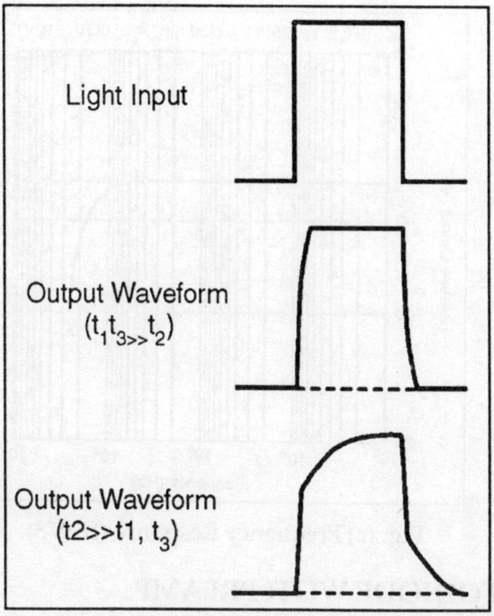

Fig. (a) Photodiode Response Waveform Example

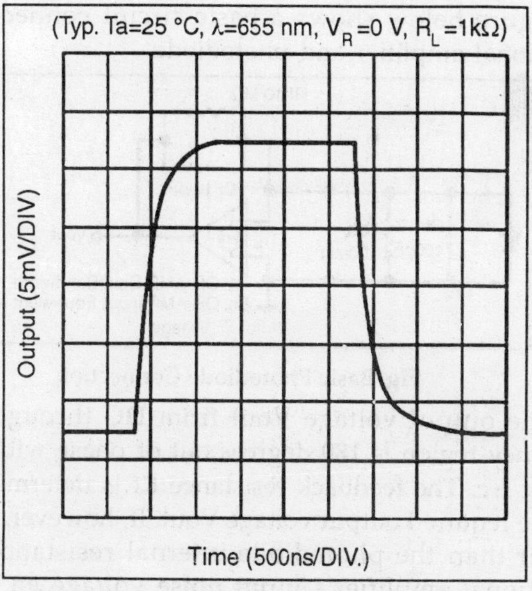

Fig. (b) Response Waveform (S2386-18K)

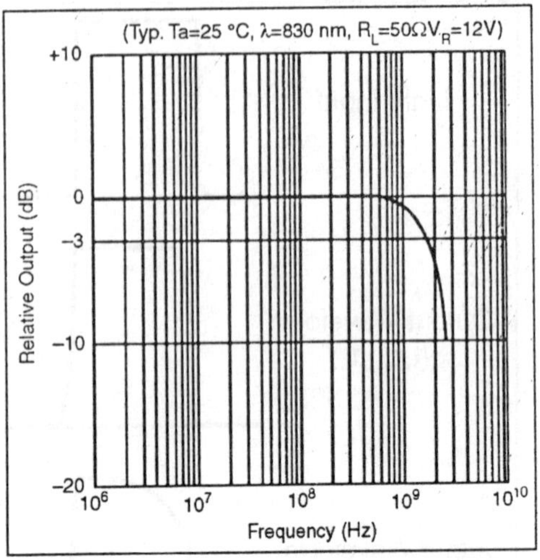

Fig. (c) Frequency Response (S5973)

SI PHOTODIODE WITH PREAMP

Feedback Circuit

Figure below shows a basic circuit connection of an operational amplifier and photodiode.

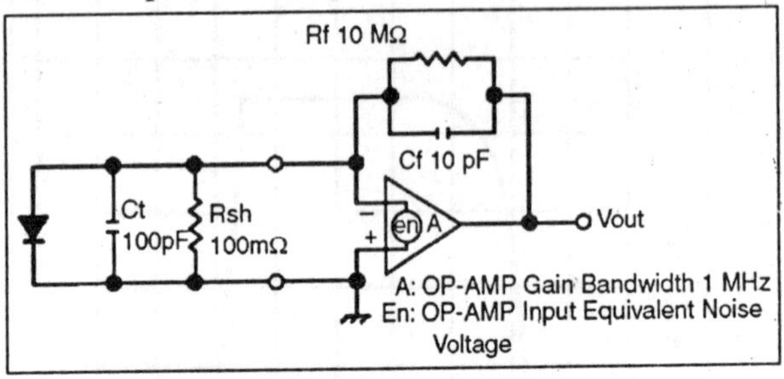

Fig. Basic Photodiode Connection

The output voltage Vout from DC through the low-frequency region is 180 degrees out of phase with the input current Isc. The feedback resistance Rf is determined by Isc and the required output voltage Vout. If, however, Rf is made greater than the photodiode internal resistance Rsh, the operational amplifier's input noise voltage en and offset voltage will be multiplied by,

$$\left(1+\frac{Rf}{Rsh}\right)$$

This is superimposed on the output voltage Vout, and the operational amplifier's bias current error will also increase. It is therefore not practical to use an infinitely large Rf. If there is an input capacitance Ct, the feedback capacitance Ct prevents high-frequency oscillations and also forms a lowpass filter with a time constant Cf × Rf value.

The value of Cf should be chosen according to the application. If the input light is similar to a discharge spark, and it is desired to integrate the amount of light, Rf can be removed so that the operational amplifier and Cf act as an integrating circuit. However, a switch is required to discharge Cf before the next integration.

Bias Current

Since the actual input impedance of an operational amplifier is not infinite, some bias current that will flows into or out of the input terminals.

This may result in error, depending upon the magnitude of the detected current. The bias current, which flows in an FET, input operational amplifier is sometimes lower than 0.1 pA. Bipolar operational amplifiers, however, have bias currents ranging from several hundred pA to several hundred nA. However, the bias current of an FET operational amplifier increases two-fold for every increase of 5 to 10 °C in temperature, whereas that of bipolar amplifiers decreases with increasing temperature.

The use of bipolar amplifiers should be considered when designing circuits for high temperature operation. As is the case with offset voltage, the error voltage attributable to the bias current can be adjusted by means of a potentiometer connected to the offset adjustment terminals. Furthermore, leakage currents on the PC board used to house the circuit may be greater than the operational amplifier's bias current. Consideration must be given to the circuit pattern design and parts layout, as well as the use of Teflon terminals and guard rings.

Gain Peaking

The frequency response of a photodiode and operational amplifier circuit is determined by the time constant Rf × Cf. However, for large values of terminal capacitance (i.e. input capacitance) a phenomenon known as gain peaking will occur. Figure below shows an example of such a frequency response. It can be seen from the figure that the output voltage increases sharply in the high frequency region, causing significant ringing

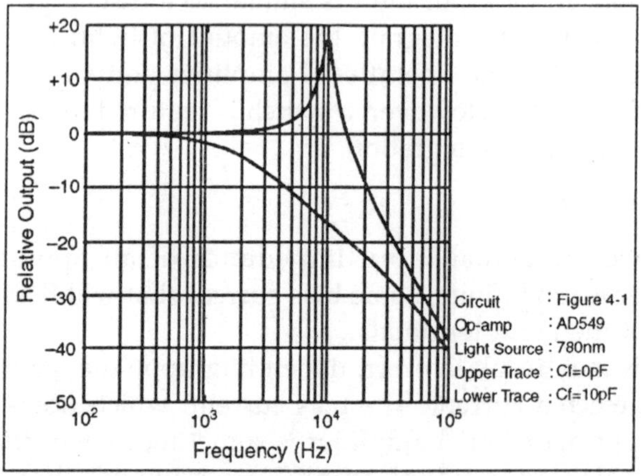

Fig. (a) Frequency response

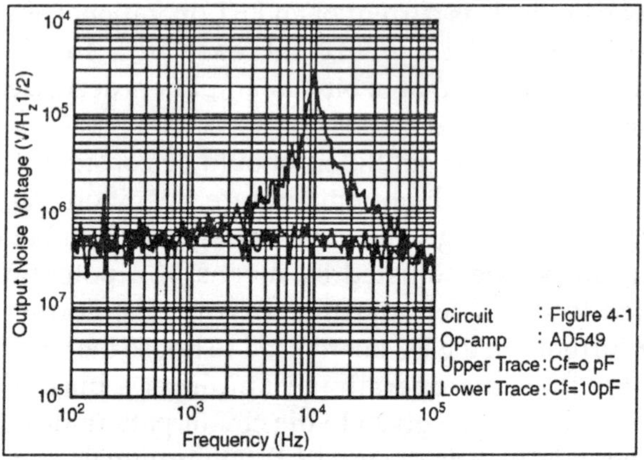

Fig. (b) Frequency response of noise output

[See the upper trace in (a).]

in the output voltage waveform in response to the pulsed light input. This gain operates in the same manner with respect to operational amplifier input noise and may result in abnormally high noise levels.

[See the upper trace in fig (b).]

This occurs at the high frequency region when the reactance of the input capacitance and the feedback capacitance of the operational amplifier circuit jointly form an unstable amplifier with respect to input amplifier noise. In such a case, loss of measurement accuracy may result.

Gain Peaking Elimination

To achieve a wide frequency characteristic without gain peaking and ringing phenomena, it is necessary to select the optimum relationship between the photodiode, opera-tional amplifier and feedback element.

It will prove effective in the case of photodiodes to reduce the terminal capaci-tance Ct, as was previously explained in the section on Re-sponse speed and frequency response.

In the operational amplifier, the higher the speed and the wider the band-width, the less the gain peaking that occurs. However, if adequate internal phase compensation is not provided, os-cillation may be generated as a result.

A feedback element, not only the resistance but also the feedback capacitance should be connected in parallel, as explained previously, in order to avoid gain peaking. The gain peaking phe-nomena can be explained as follows, using the circuit shown in Figure above.

As shown in Figure below, the circuit gain of the operational amplifier is determined for the low-frequency region simply by the resistance ratio of Rsh to Rf.

From the frequency,

$$f1 = \frac{Rsh + Rf}{2\pi RshRf(cf + Ct)}$$

gain begins to increase with frequency as shown in re-gion (2).

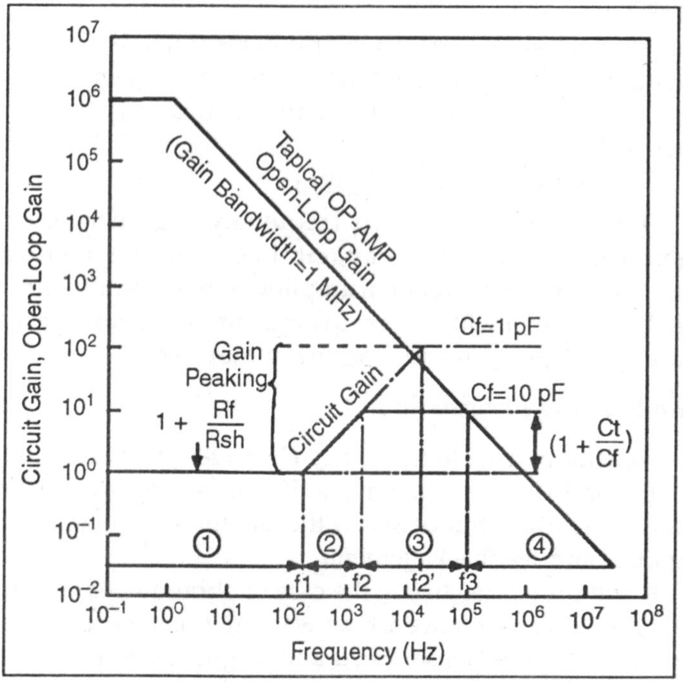

Fig. Graphical Representation of Gain Peaking

Next, at the frequency,

$$f2 = \frac{1}{2\pi CfRf}$$

and above, the circuit gain of the operational amplifier enters a flat region (region (3)) which is determined by the ratio of Ct and Cf. At the point where frequency f3 intersects the open-loop gain frequency response at rolloff (6 dB/octave) of the operational amplifier, region is entered. In this example, f1 and f2 correspond to 160 Hz and 1.6 kHz respectively under the conditions of Figure 4-1. If Cf is made 1 pF, f2 shifts to f2′ and circuit gain increases further. What should be noted here is that, since the setting of increasing circuit gain in region *f* exceeds the open-loop gain curve, region (3) actually does not exist. As a result, ringing occurs in the pulsed light response of the operational amplifier circuit, and the gain peaking occurs in the frequency, then instability results.

SI APD

Benefits of APD

When using a opto-semiconductor for low-light-level measurement, it is necessary to take overall performance into account, including not only the opto-semiconductor characteristics but also the readout circuit (operational amplifier, etc.) noise.

When a Si photodiode is used as a photodetector, the lowest detection limit is usually determined by the readout circuit noise because photodiode noise level is very low. This tendency becomes more obvious when the higher frequency of signal to be detected.

This is because the high-speed readout circuit usually exhibits larger noise, resulting in a predominant source of noise in the entire circuit system.In such cases, if the detector itself has an internal gain mechanism and if the output signal from the detector is thus adequately amplified, the readout circuit can be operated so that its noise contribution is minimized to levels equal to one divided by gain (1/10 th to 1/100 th).In this way, when the lowest detection limit is determined by the readout circuit, use of an APD offers the advantage that the lowest detection limit can be improved by the APD gain factor to a level 1/10 th to 1/100 th of the lowest detection limit obtained with normal photodiodes.

Noise Characteristic of APD

When the signal is amplified, the inherent excess noise resulting from statistical current fluctuation current fluctuation in the avalanche multiplication process is also generated. This noise current can be expressed by the following equation:

$$in = \sqrt{2qILM^2FB} \text{.....}(5-1)$$

In the range of M=10 to 100, F is approximated M^x.

(*F*: Excess noise factor, M:Gain, IL: Photocurrentat at M=1), (*q*: Electron charge, B: Bandwidth, x: Excess noise index)

In PIN photodiodes, using a large load resistance is not practical since it limits the response speed, so the circuit noise is usually dominated by the thermal noise of the photodiode. In contrast, the gain of an APD, which is internally amplified, can be increased until the shot noise reaches the same level as the thermal noise.

The APD can therefore offer an improved S/N without impairing the response speed.

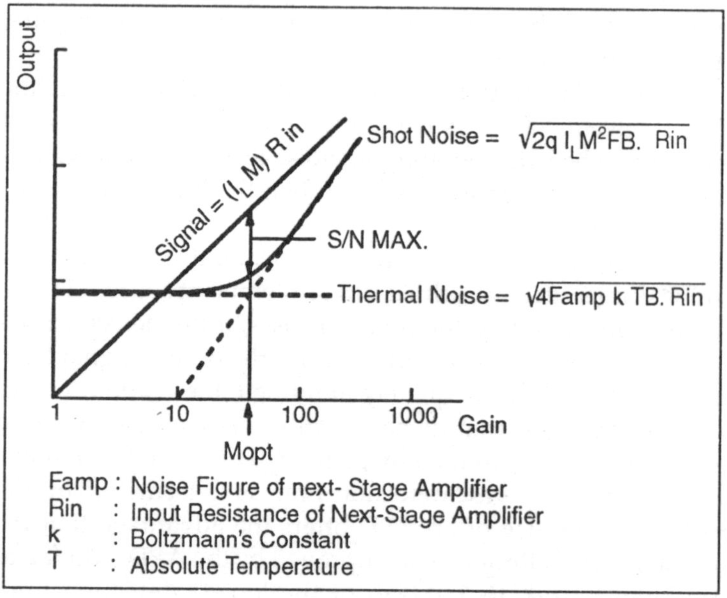

Fig. Noise Characteristic of APD

Spectral Response of APD

The spectral response characteristics of the APD are almost the same as those of normal photodiodes if a bias voltage is not applied. When a bias voltage is applied, the spectral response curve will change.

This means that the gain changes depending on the incident light wavelength. This is because the penetration depth of light into the silicon substrate depends on the wavelength so that the wavelength absorption efficiency in the light absorption region differs depending on the APD structure.

To allow selection of spectral response characteristics,

Hamamatsu provides two types of Si APDs: S2381 series and S6045 series for near infrared detection and S5343 series for light detection at shorter wavelengths. Figure below shows typical spectral response characteristics measured with a gain of 30 at 650 nm wavelength.

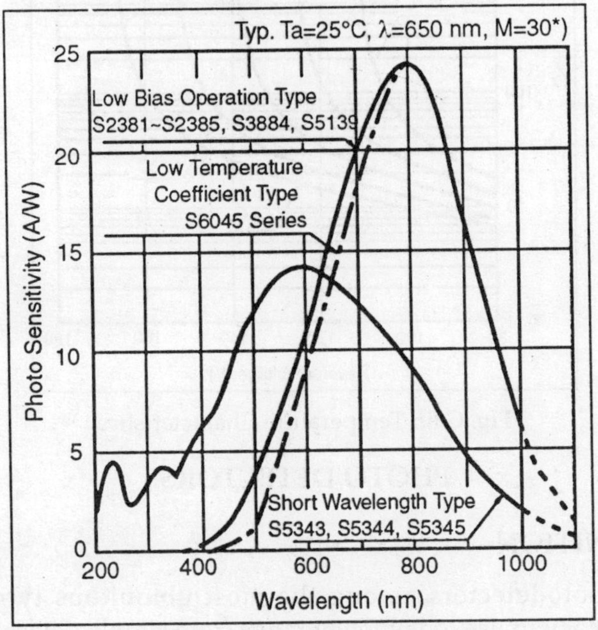

Fig. Spectral Response

Temperature Characteristic of Gain

APD gain varies with temperature. For example, when an APD is operated at a constant bias voltage, the gain decreases with increasing temperature.

Therefore, in order to obtain a constant output, it is necessary to vary the bias voltage according to the APD temperature or to keep the APD at a constant temperature.In S2381 series, the temperature coefficient of the bias voltage is nearly equal to that of the breakdown voltage which is 0.65 V/°C Typ. at a gain of 100.Hamamatsu also provides S6045 series APDs which are designed to have an improved temperature coefficient (0.4 V/°C Typ.).

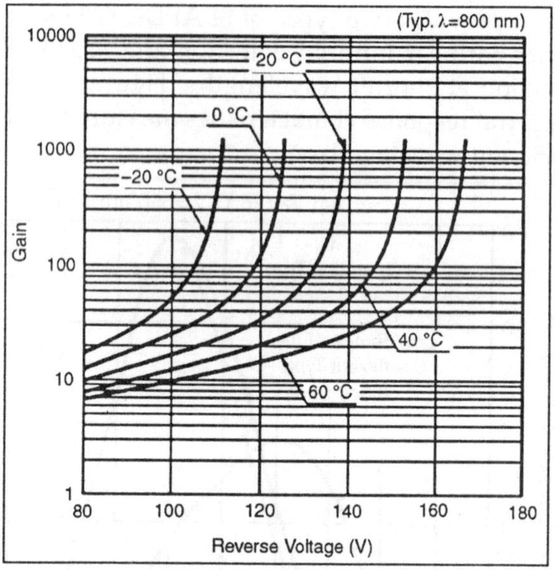

Fig. Gain Temperature Characteristics

PHOTO DETECTORS

DEFINITION

Photodetectors among the most ubiquitous types of technology in use today. They range from simple devices that automatically open supermarket doors, to receivers on TV and VCR remote controls, to photodiodes in a fiberoptic connection, to the CCD in a video camera, to enormous arrays used by astronomers to detect radiation from the other side of the universe. Photodetectors are present in a huge variety of devices used in commerce, industry, entertainment and research. In fact, the field of photodetector design and use has grown to the point that few practitioners have a complete overview.

For our purposes, photodetectors include any device for registering photons with frequencies above that of radio waves—from far infrared on up to gamma rays. In this article we survey the main types of applications that use photodetectors. Broadly speaking, these applications fall into two general groupings—communications and remote sensing. In communications, the radiation is simply the carrier for an encoded signal, while in the various forms of remote sensing, the radiation is the signal, conveying information about an object or scene.

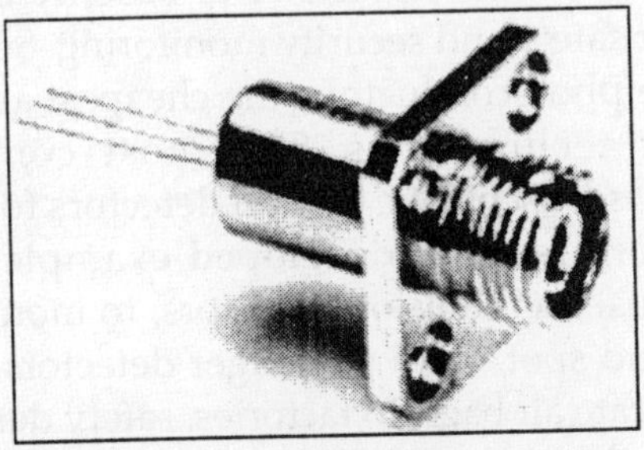

Fig. Galnas Pin Detectors

Perhaps the most ubiquitous communications market is fiberoptic communications systems, in which photodetectors, generally operating in the infrared (IR), pick up high-speed signals. These detectors do not need high sensitivity, since the laser drivers provide plenty of radiation to the fibre, but they must have extremely fast response, exhibit high reliability, and have a low cost.

Photodiodes, especially those based on indium gallium arsenide (InGaAs) are the workhorses of optical communications (see Fig. 1), currently achieving data communication rates as high as 2.5 Gbits/s, which is more than 200,000 times the capacity of a single copper telephone wire.

While the detectors used in fiberoptic communications are invisible to the telephone user, those used in common remote control devices are obvious and are found in nearly every American home.

Again, photodiodes sensitive to IR wavelengths are the standard, but requirements are far laxer than for fiberoptic devices, since data transmission rates are low, and only a small amount of information is generally conveyed—simple commands to change channels, or switch a VCR or TV on or off. Even remote controls, however, are increasing in sophistication—cordless mouse devices for PCs, for example, use detectors that measure the relative strength of signals to determine mouse orientation and position.

SAFETY AND SECURITY

The simplest types of remote sensing applications involve just the detection of the presence or absence of an object or a condition for safety and security monitoring. Such applications typically use photoconductors, the cheapest and most rugged of detector technologies. The most common of such applications is IR-sensitive motion detectors for home security systems. More recently developed examples include such automotive uses as collision detectors, to monitor objects in a driver's "blind spot," and passenger detectors that determine when to activate air bags. In factories, safety detectors sensitive to visible or ultraviolet (UV) wavelengths perform such tasks as electrical arc detection, automatically cutting off the current where arcing occurs.

PROCESS CONTROL

The next step up in sophistication comes in process control applications, another large volume consumer application of photodetectors. In many cases, these devices may be as simple as position sensors to check that a work piece is in the proper place or to provide feedback for robotic systems (see Fig. 2). But other systems need accurate comparison of radiation intensity at different wavelengths, generally requiring accurately repeatable detectors such as photodiodes or phototransistors. In recycling plants, for example, the natural fluorescence of certain plastics can be used with optically filtered detectors to sort different materials, while other spectroscopic tests can differentiate different types of glass.

Fig. Germanium Position Sensors

Comparison of radiation intensity at two or more IR wavelengths is used in pyrometry to remotely measure and control high-temperature processes. Similar techniques in the UV are used to monitor and control flames. In some of these applications, in which radiation is intense, ruggedized bolometers—which measure radiation by the heat transferred, rather than by quantum detection of individual photons—are frequently preferred. At the end of the production process, detectors are used as part of nondestructive testing units, using thermal scans to check for physical defects or for surface composition and finish. Increasingly, imaging systems, especially CCDs, are used in machine vision systems that are incorporated into more flexible robots, not only in factories but also, to a limited extent, in mobile devices used in service institutions, such as hospitals.

Process control applications in factories often require detector systems that can endure extreme environmental conditions. In the food industry, for example, sanitizing caustic solutions are sprayed out at high temperature and at pressures up to 1200 psi, necessitating extremely durable photodetector packaging. In metalworking plants, detectors may require Teflon housing to facilitate removal of molten metal splatter from spot welding. Such environments also put a premium on detector materials able to withstand wide ranges in temperature conditions to reduce insulation needs.

A third high-volume detector application is in all forms of video cameras, from the large cameras used by broadcast and cable TV studios to a palm-sized camcorder. While this is a relatively mature field, refinements in light sensitivity, response time, and resolution are continuing in the basic CCD technology that dominates the video camera business (see Fig. 3). The latest development in the field involves the extension of mass-production video camera technology to IR night-vision applications that were previously limited mostly to defence uses.

ADVANTAGE OF DETECTOR

The most advanced detectors tend to be made for low-volume specialized fields, such as laboratory research, biomedicine, defence, environmental monitoring, and astronomy. Thermography, measuring the heat radiated by the human body, has long been a routine diagnostic test, requiring IR detectors. Recent developments have allowed research biologists to observe rapid biochemical reactions with ultrahigh-speed CCDs. Cameras used for tracing chemical diffusion in individual cells, for example, typically have read-out rates of several megahertz and frame rates of up to 10,000 per second. In spectroscopic applications, the high sensitivity of photomultiplier tubes (PMTs) makes them the workhorse of laboratory detectors.

Undoubtedly the biggest drivers of high-technology detectors, especially in the IR, have been defence applications. For decades, the armed forces have used IR detectors on heat seeking missiles, thus setting off a long-term arms race, as decoys such as magnesium flares confused these detectors, prompting the development of more sophisticated devices to distinguish the decoys from real targets. This has led to the development of tunable IR detectors that can pick out jet exhaust spectra, satellite-borne staring array detectors to isolate warm targets against a warm background, and continuous improvements in IR video cameras for night vision. Infrared scanning from aircraft reveals the location of land mines, which show up as regions of warmer soil. The ultimately futile missile defence Programme, the Strategic Defence Initiative (SDI), while it never arrived at a practical way to shoot down ICBMs, did lead to the development of sensitive far-IR detectors for wavelengths longer than 5 μm, which subsequently found use in astronomy, among other places.

With the scaling back of SDI and the end of the Cold War, many military technologies are moving out into civilian applications. Work is well advanced on adapting military night vision IR detector systems to enhance driver vision beyond the reach of headlights. Workhorses of low-light detection such as avalanche photodiodes first emerged from Pentagon-funded research.

USES

Environmental Sensing

Environmental monitoring today uses a broad range of photodetectors from the UV to the IR. Typically, signals are low intensity, so the primary detectors are PMTs and avalanche photodiodes (solid-state photomultipliers). Pollution detection generally relies on UV spectroscopy, with detectors measuring the strength of absorption lines for such pollutants as sulfur dioxide, nitrous oxides, and ozone.

Fluorescence spectroscopy allows detection of extremely small amounts of pollutants such as benzene, toluene, xylene and sulfur oxides, which can be measured at a parts-per-billion level.

Another sensitive technique is chemiluminescence, in which a reactant is introduced that selectively combines with a pollutant, producing a faint luminescence in the process. Large active areas and low noise levels are required for detectors in such applications.

More straightforward techniques are used for monitoring solid particulates in air and water. Here, the amount of light scattered by the particles is a good continuous measure of pollution levels. Ample light is often available, so photodiodes can be used, taking advantage of their robustness and fast response times.

For monitoring of pollution over large areas, lidar scattering techniques are increasingly used. In this method, a laser sends out light pulses at a repetition rate of about 50 Hz; these pulses scatter off pollutants in the air. Sensitive detectors, generally PMTs, record the amount of light scattered and the time of arrival of the scattered light. In this way, the distance to the pollutant can be measured. As the lidar scans across a region, the distribution of a given pollutant can be mapped.

Space-based environmental monitoring, both for tracking pollutants and for weather and climate analysis make greater demands on detectors, both for sensitivity and for compactness.

Satellite-mounted IR spectrometers can measure ozone

concentrations from space, sometimes using pyroelectric detectors to convert heat into electric impulses. Space-based IR detectors for temperature monitoring are generally mercury-cadmium-telluride IR arrays that were first developed for defence purposes.

Astronomy

Without doubt the most challenging applications for photodetectors come in the field of astronomy, where the range of wavelengths studied extends from the far-IR at hundreds of microns to cosmic ray photons with 1020 eV of energy and wavelengths of 10-20 μm, a range of 22 orders of magnitude. At the long wavelength end, the European Space agency is making plans for FIRST, the far-infrared and submillimeter space telescope, which will open up the one spectral band not yet surveyed—from 60 to 670 μm.

The design will use arrays of germanium gallium photoconductors for 100 to 200 μm and arrays of bolometers cooled to 0.1 K for the rest of the band.

At near-IR wavelengths, focal plane arrays of indium antinomide, mercury cadmium telluride, and platinum silicide, all initially developed for the military, are becoming routing accessories for both ground- and space-based telescopes.

Miniaturized IR and CCD cameras on board the spacecraft Clementine enabled it to survey the moon at wavelengths ranging from the UV to 9.5 μm in the far-IR.

Fig.3 Ultraminiature, High-resolution CCD

At UV wavelengths, necessarily viewed from space because the atmosphere is mostly opaque to UV, advances in microchannel-plate PMTs have led to major improvements in sensitivity, evident in such instruments as those used in the ASTRO platform on board the space shuttles.

At still higher energies, x-ray telescopes such as ROSAT, launched by the European Space Agency, are opening new windows into the most energetic processes in the universe. X-ray detectors generally are based on approaches different from those used for longer wavelength radiation.

In gas-filled detectors, electrons are knocked out by x-rays, and electron multiplication occurs in the gas as electrons avalanche toward a high-voltage wire.

However, techniques borrowed from other parts of the spectrum, such as microchannel plates for electron multiplication, can be used by x-ray detectors as well.

For soft x-rays, below 10 keV, specially designed CCDs are finding greater application. Higher energy detectors often employ calorimeters, which measure the total heat liberated when x-rays are absorbed. And research is currently developing fundamentally new types of detectors based on superconducting tunnel junctions.

TYPES

Photovoltaic Effect Detector

The *photovoltaic effect* occurs at a junction in a semiconductor. The junction is the boundary between a region where the conductivity is due to electrons and a region where the conductivity is due to holes (the absence of electrons). This is called a pn junction.

At the junction, an electric field is present internally because there is a change in the level of the conduction and valence bands.

This change leads to the familiar electrical rectification effect produced by such junctions. The photovoltaic effect is the generation of a voltage when light strikes a semiconductor pn junction.

The photovoltaic effect is measured using a high-impedance voltage-measuring device, which essentially measures the open-circuit voltage produced at the junction.

In the dark, no open circuit voltage is present. When light falls on the junction, the light is absorbed and, if the photon energy is large enough, it produces free hole-electron pairs. The electric field at the junction separates the pair and moves the electron into the n-type region and the hole into the p-type region.

This leads to an open circuit voltage that can be measured externally. This process is the origin of the photovoltaic effect. We note that the open-circuit voltage generated in the photovoltaic effect may be detected directly and that no bias voltage nor load resistor is required.

Photoemissive Effect Detector

Now we turn to the *photoemissive effect*. The photoemissive effect involves the emission of electrons from a surface irradiated by quanta of light energy.

A photoemissive detector has a cathode coated with a material that emits electrons when light of wavelength shorter than the cutoff wavelength falls on the surface.

The electrons emitted from the surface are accelerated by a voltage to an anode, where they produce a current in an external circuit. The detectors are enclosed in a vacuum environment to allow a free flow of electrons from anode to cathode. Some spectral response curves for photoemissive cathodes are shown in Figure below. The cathodes are often mixtures containing alkali metals, such as sodium and potassium, from which electrons can easily be emitted. The responsivity in mA/watt of these devices is shown in the figure from the ultraviolet to the near infrared.

Curve 1 is the response of a bialkali type of cathode with a sapphire window; curve 2 is for a different bialkali cathode with a lime glass window; curve 3 is for a multialkali cathode with a lime glass window; and curve 4 is for a GaAs cathode with a 9741 glass window. The curves labeled 1% and 10% denote what the response would be at the indicated value of quantum efficiency.

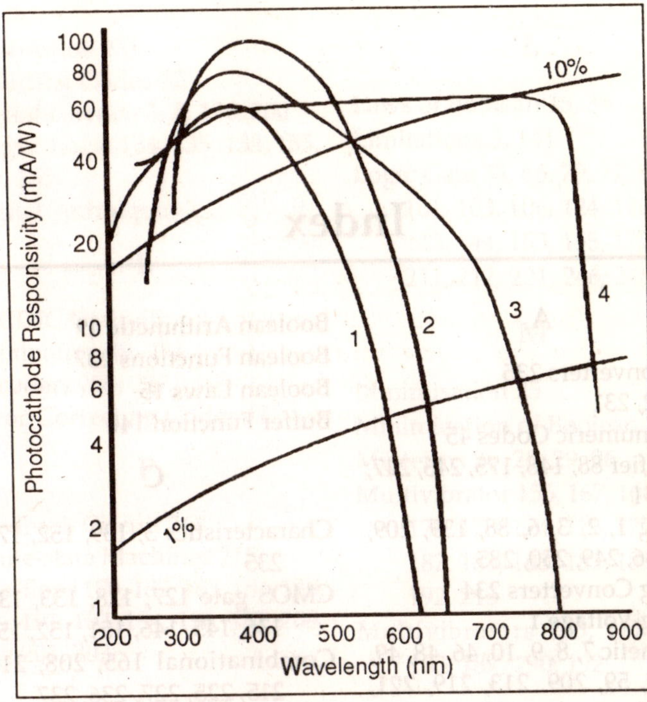

Fig. Respon se as a Function of Wavelength for a Number of Photoemissive Surfaces.

S

T

V

W